Protect Your Family and Secure Your Home During Any Disaster

Essential Bug-in Survival Guide for Homes

Complete Practical Strategies and Projects for Self-Reliance, Grid-Down Living and Family Preparedness for Any Emergency

By Scott Barnabas

The Complete Emergency Survival Handbook for Every Home

Copyright © 2024 by Scott Barnabas

All Rights Reserved.

No part of this book may be reproduced, stored in a retrieval system, or transmitted in any form or by any means, electronic, mechanical, photocopying, recording, or otherwise, without the prior written permission of the publisher, except for the use of brief quotations in a review.

Disclaimer: This book is intended for informational purposes only. The author and publisher make no representations or warranties regarding the accuracy, applicability, or completeness of the content. The information contained herein is based on the author's experience and research, but readers are advised to consult appropriate professionals for specific advice tailored to their circumstances. The publisher and author shall not be held liable for any damages resulting from the use or misuse of the information in this book.

Trademarks: All trademarks, registered trademarks, and service marks used in this book are the property of their respective owners.

First Edition: 2024

Table of Contents

FOREWORD — 7
1. INTRODUCTION — 9
   - What It Means to Bug In — 9
   - How to Use This Guide — 10
2. QUICK-START GUIDE FOR BEGINNERS — 12
   - The 72-Hour Survival Plan — 12
   - Essentials Checklist for Immediate Readiness — 15
   - Common Mistakes to Avoid — 19
3. UNDERSTANDING THREATS AND RISKS — 23
   - Identifying Local and Global Emergencies — 23
   - Home Vulnerability Assessment — 27
   - Deciding Between Bugging In and Bugging Out — 31
4. CREATING A FAMILY-CENTERED BUG-IN PLAN — 35
   - Building a Custom Emergency Blueprint — 35
   - Communication Strategies for Loved Ones — 39
   - Caring for Pets, Children, and the Elderly — 43
5. HOME DEFENSE AND SECURITY — 47
   - Fortifying Doors, Windows, and Entry Points — 47
   - DIY Home Surveillance Systems — 50
   - Creating a Safe Room for Your Family — 54
   - Defensive Landscaping for Added Protection — 58
6. STOCKPILING AND RESOURCE MANAGEMENT — 62
   - Food Storage Essentials: Short and Long-Term — 62
   - Water Storage, Filtration, and Purification — 65
   - Rotating and Preserving Supplies — 68
   - Medical Kits and Specialized Tools — 71
7. SCENARIOS AND CRISIS-SPECIFIC PLANNING — 75
   - Surviving Natural Disasters (Floods, Earthquakes, Storms) — 75
   - Preparing for Grid-Down and Power Outages — 78
   - Handling Pandemics and Biohazards — 82
   - Protecting Your Home During Civil Unrest — 85
8. SELF-SUFFICIENCY AT HOME — 89

    Setting Up Off-Grid Energy Solutions — 89

    Water Harvesting and Filtration Systems — 93

    Indoor and Outdoor Gardening for Sustenance — 96

    Managing Waste and Sanitation in Prolonged Crises — 100

9. DIY SURVIVAL PROJECTS — 105

    Building a Rocket Stove — 105

    Constructing Rainwater Collection Systems — 109

    Creating Homemade Air Filtration Units — 112

    Crafting Emergency Lighting Solutions — 116

10. EMERGENCY MEDICAL PREPAREDNESS — 120

    Assembling a Complete First Aid Kit — 120

    Learning Basic and Advanced Medical Skills — 123

    Managing Chronic Conditions Without External Help — 127

    Handling Psychological Stress and Trauma — 131

11. STAYING CONNECTED AND INFORMED — 135

    Accessing Reliable News and Alerts — 135

    Using Backup Communication Tools and Radios — 139

    Coordinating with Neighbors and Networks — 143

12. FIRE SAFETY AND HOME MANAGEMENT — 145

    Preventing House Fires During Crisis Scenarios — 145

    Cooking and Heating Alternatives Without Power — 147

    Emergency Fire Suppression Techniques — 151

13. FINANCIAL AND LEGAL READINESS — 155

    Securing Critical Documents in a Crisis — 155

    Creating a Financial Emergency Fund — 158

    Preparing for Economic Disruptions — 162

14. LIVING THROUGH LONG-TERM CRISES — 165

    Sustaining Morale and Motivation During Economic Disruptions — 165

    Adjusting to a New Normal During Economic Disruptions — 166

    Teaching Survival Skills to Children — 169

15. CONCLUSION — 173

    The Importance of Ongoing Preparedness — 173

    Building a Lifestyle of Readiness and Resilience — 175

APPENDICES 177
  Comprehensive Supply Checklists 177
  Resource Lists for Preppers 182
  DIY Templates for Home Projects 187
  Emergency Plan Examples 192

FOREWORD

In the heart of the Alaskan wilderness, a man named Thompson found himself in the middle of one of the most intense challenges of his life. It was the dead of winter, and the weather had turned against him. The massive blizzard that slammed into his remote cabin left him and his supplies completely cut off from the outside world. With no way to get in or out, no cell service, and the bitter cold creeping in Thompson's survival instincts kicked in.

Thompson wasn't just any survivalist. He had spent years learning how to live off the land, how to endure in extreme conditions, and how to adapt when everything around him seemed to be working against him. But this was different. This storm wasn't just a typical Alaskan winter squall. It was fierce, unrelenting, and it wasn't going to let up. Every moment mattered. And the clock was ticking.

As the snow piled up around his cabin, Thompson knew he had a limited window of time before his situation could become worst. The first challenge was securing his shelter. Thompson had built his cabin with durability in mind, but he had never imagined a storm of this magnitude. The winds howled, whipping the snow into a frenzy, and he could hear the creaking of the wood as the structure strained against the pressure. It was a constant reminder that nature was not to be underestimated.

Thompson wasted no time reinforcing the walls and windows, using every tool at his disposal. He even fashioned a makeshift barrier out of snow and ice to prevent the wind from seeping in. Inside, the temperature was dropping rapidly, and he needed to act fast to avoid freezing. He had a small wood-burning stove, but firewood was limited, and he couldn't afford to burn through his supply too quickly. So, he rationed it carefully, using it only when necessary and supplementing it with other means of generating heat.

For the next few days, Thompson relied on his skills and knowledge to survive. The food supplies were sufficient for a while, but he couldn't just rely on canned goods and dried food. He knew how to hunt, fish, and forage, but with the storm still raging, going outside was out of the question. Instead, he turned his attention to preserving what he had, making sure to ration each meal meticulously. It was about more than just surviving the immediate threat—it was about ensuring he had enough to last until the storm passed.

The storm didn't stop for days. As Thompson faced one challenge after another—keeping his shelter intact, preserving his food, and staying warm—he drew on his training and preparation. But it wasn't just about the physical resources he had—it was his mindset, the resilience that had kept him going through every trial, that made the difference. He wasn't afraid of the storm. He had seen worse. What worried him most was the unknown—the possibility that this crisis could drag on far longer than expected, and that the resources he had would eventually run out.

On the fifth day, as the snowstorm finally started to ease, Thompson had a moment of clarity. He wasn't just surviving. He was thriving. Because he had planned. He had prepared. And he had the mental fortitude to face whatever came his way. When the storm eventually cleared, Jack was the first to step outside, taking in a world that had been transformed by nature's fury. But he knew, more than ever, that he was ready for whatever might come next. He had weathered the storm, both physically and mentally, and emerged stronger for it.

Thompson's story is not just one of survival—it's one of preparedness, resilience, and mastery of the elements. It shows us what it truly means to be self-sufficient and ready for anything life might throw at us. This is exactly what this book Essential Bug-In Survival Guide for Homes aims to equip you with: the skills, strategies, and mindset you need to not only survive but thrive in the face of any crisis.

This book is designed to bring you real-world applications of bug-in survival strategies that can be adapted for civilian life. Whether you're in a remote cabin, a suburban home, or an urban apartment, you will find practical, easy-to-follow tips for home preparedness that will help you face everything from power outages to societal collapse. The guide includes a step-by-step approach, complete with checklists, visuals, and actionable advice, to help you take control of your home, your family's safety, and your future.

Just as Thompson had the confidence to face the worst that nature could throw at him, this guide will help you gain the same confidence. You'll learn how to prepare your home, how to defend it against any threat, and how to make sure your family is safe, secure, and self-sufficient. It's not about fearing the worst. It's about being ready for anything.

As you read through these pages, remember Thompson's story. His survival wasn't based on luck—it was based on preparation. This guide is here to ensure you're not just ready for a crisis; you're ready to take on the world, no matter what it throws at you.

~ Scott Barnabas

1. INTRODUCTION

What It Means to Bug In

"Bugging in" refers to the strategy of staying at your current location—typically your home—during an emergency or disaster, rather than evacuating to another site. This approach involves preparing your residence to serve as a self-sustaining refuge, equipped to handle various crises such as natural disasters, societal unrest, or infrastructure failures.

Key Aspects of Bugging In

- **Shelter and Security:** Your home becomes a fortress, providing protection from external threats. This includes reinforcing entry points, securing windows and doors, and establishing a safe room if necessary.
- **Resource Management:** Stockpiling essential supplies is crucial. This encompasses food, water, medical supplies, and other necessities to sustain you and your family for an extended period.
- **Self-Sufficiency:** Developing skills and systems to generate power, purify water, and grow food can reduce reliance on external sources, which may be disrupted during a crisis.
- **Communication:** Establishing reliable communication methods, such as battery-powered radios or satellite phones, ensures you can receive updates and coordinate with others when traditional networks are unavailable.

Benefits of Bugging In

- **Familiar Environment:** Staying in a known setting can reduce stress and provide comfort during uncertain times.
- **Community Support:** Remaining in your neighborhood allows for mutual assistance with neighbors, fostering a sense of solidarity.
- **Resource Availability:** Your home likely contains tools, materials, and resources that can be utilized for survival purposes.
- **Challenges to Consider**
- **Limited Space:** Urban and suburban homes may have constraints on storage and space for self-sufficiency projects.
- **Security Risks:** Increased threats from looting or civil unrest may necessitate enhanced security measures.
- **Resource Management:** Ensuring a sustainable supply of essentials like food and water requires careful planning and regular monitoring.

Bugging in is a proactive strategy that requires thorough preparation and adaptability. By transforming your home into a well-equipped, self-sustaining environment, you can enhance your resilience and ability to navigate various emergencies effectively.

How to Use This Guide

Welcome to the Essential Bug-In Survival Guide for Homes. This comprehensive resource is designed to equip you with the knowledge and practical skills necessary to transform your home into a self-sustaining sanctuary during emergencies.

1. Understand the Structure

The guide is organized into clear, actionable sections, each focusing on a critical aspect of home preparedness:

Introduction: Provides an overview of the importance of bugging in and sets the stage for the strategies discussed.

Assessing Your Home: Guides you through evaluating your current living situation to identify strengths and areas for improvement.

Essential Supplies: Details the must-have items for food, water, medical needs, and other essentials.

Home Fortification: Offers step-by-step instructions on securing your home against potential threats.

Energy Independence: Explores alternative energy solutions to maintain power during outages.

Health and Sanitation: Discusses maintaining hygiene and health standards in a crisis.

Mental Resilience: Provides strategies for psychological preparedness and stress management.

2. Tailor the Content to Your Needs

While the guide offers general advice, it's essential to adapt the recommendations to your specific circumstances:

Assess Your Environment: Consider your home's location, size, and the typical risks associated with your area.

Inventory Your Resources: Take stock of your current supplies and skills to identify gaps.

Set Priorities: Focus on areas that require immediate attention based on your assessment.

3. Implement Practical Steps

Each chapter includes actionable steps, checklists, and visuals to guide you:

Checklists: Use these to track your progress and ensure nothing is overlooked.

Step-by-Step Instructions: Follow detailed procedures for tasks like building a shelter or purifying water.

4. Engage in Continuous Learning

Preparedness is an ongoing process:

Practice Skills: Regularly rehearse critical skills to build confidence and proficiency.

Stay Informed: Keep abreast of new developments in survival strategies and home preparedness.

Join Communities: Engage with local or online preparedness groups to exchange knowledge and experiences.

5. Review and Update Regularly

As circumstances change, so should your preparedness plan:

Regular Assessments: Periodically evaluate your supplies and home security measures.

Adapt to New Threats: Stay informed about emerging risks and adjust your plans accordingly.

Maintain Flexibility: Be prepared to modify your strategies as new information and resources become available.

By actively engaging with this guide and integrating its strategies into your daily life, you can enhance your resilience and ensure that your home remains a safe haven during times of crisis.

2. QUICK-START GUIDE FOR BEGINNERS

The 72-Hour Survival Plan

The first 72 hours of a disaster or emergency are often the most critical. Whether it's a natural disaster, a power grid failure, or an unexpected crisis, the initial three days will determine how well you can manage your situation, stay safe, and preserve your resources. This plan is designed to guide you through those first 72 hours, ensuring that you have the essentials covered and can make it through the critical early period.

1. Immediate Response: Assess Your Situation (First 6 Hours)

In the first few hours after a crisis hits, the main goal is to assess your immediate environment and ensure the safety of your family. Prioritize these steps:

Ensure Safety

Secure your environment: Check for hazards such as fires, gas leaks, structural damage, or electrical issues. If necessary, evacuate your home if it's unsafe.

Account for everyone: Ensure all family members, pets, and anyone in your care are safe and accounted for.

Establish communication: Try to contact loved ones, neighbors, or local emergency services to get updated information.

Establish a Safe Room or Meeting Area

Choose a secure location: If you are indoors, designate a safe room where you can shelter from potential threats. A basement, interior room, or any space without windows and with multiple exit routes is ideal.

Gather emergency gear: Bring your emergency kits, medications, first-aid supplies, flashlights, and essential tools into the safe space.

2. Stabilize Your Resources (6–24 Hours)

Once you've secured your immediate environment, the next step is to stabilize your resources and begin your self-sufficiency strategies. In this period, it's important to ration and manage your supplies efficiently.

Water and Food

Water supply: If water service has been disrupted, use your stored water or gather from any available, safe sources (e.g., swimming pools, bathtubs, or melted snow). Begin rationing water immediately.

Food supply: Use your stored food or any non-perishable items you have. Opt for high-calorie, easy-to-prepare meals to conserve energy and resources.

Power and Heat

Power: If the grid is down, assess your backup power sources. If you have a generator, ensure it's running safely. Limit power usage to essential items like lights, communication devices, and small appliances.

Heat and warmth: If it's cold and heating systems are down, layer up with clothing, use blankets, and consider alternative heat sources like candles or wood stoves (ensure proper ventilation).

3. Secure and Fortify Your Home (24–48 Hours)

With food, water, and shelter secured, your next priority is reinforcing your home to protect against any external threats, such as looting or invasion. The ability to fortify your space will buy you time until help arrives or conditions stabilize.

Home Fortification

Lock doors and windows: Ensure all doors are securely locked, and windows are closed and covered. If necessary, board up windows or use furniture to block entry points.

Create barricades: If you anticipate external threats, barricade weaker areas, particularly entry points like doors and windows. Use furniture, heavy objects, or anything that could provide additional protection.

Establish a watch system: If you're in a neighborhood or urban environment, maintain a watch system with neighbors, rotating shifts to observe any suspicious activity.

Weaponry and Self-Defense

Arm yourself safely: If you have firearms, ensure they are readily accessible and you know how to use them. If firearms are not an option, consider other self-defense tools like knives or pepper spray.

Create a defense strategy: Plan for potential scenarios, including how you would defend yourself or your family if intruders or hostile forces were to approach.

4. Communication and Information (48–72 Hours)

By this time, you should have stabilized your position, secured your family, and fortified your home. The next step is to maintain communication with the outside world and gather critical information.

Emergency Communication

Use a battery-powered or hand-crank radio: If the power is down, use a hand-crank or battery-powered radio to listen for local emergency broadcasts, weather updates, or government announcements.

Preserve phone battery: Conserve your phone battery and use it only for critical communication. Keep your devices charged when possible and limit non-essential usage.

Assess the Situation

Monitor news and updates: Track the progression of the crisis. If there's a long-term disruption, begin planning for the next steps beyond the initial 72 hours. This could include preparing for longer-term self-sufficiency or evacuation if necessary.

Stay Calm and Stay Informed

Mental preparedness: It's easy to become overwhelmed, especially in the initial stages of a crisis. Stay focused, and take regular breaks to evaluate your situation with a clear head. Being well-informed is key to making decisions that will help ensure your survival in the long run.

5. Transition to Long-Term Survival (Beyond 72 Hours)

Once the initial 72-hour period has passed, you will have a clearer picture of the ongoing situation and the ability to make informed decisions. Use the following guidelines to transition into long-term preparedness:

Reassess your resources: Take stock of your food, water, and medical supplies. Adjust your usage to ensure sustainability over the long term.

Community support: If the situation persists, reach out to your local community for mutual aid. Sharing resources, knowledge, and labor can help everyone in your neighborhood survive.

Emergency plans: Consider alternate evacuation routes if the situation worsens. Have backup plans for relocating or seeking external help if necessary.

The 72-Hour Survival Plan serves as a blueprint for effectively managing the immediate aftermath of a disaster. By staying calm, rationing your resources, and ensuring that you're well-prepared, you give yourself the best chance to navigate the chaos of an emergency. Always remember that the initial three days are just the beginning — long-term survival requires a constant commitment to preparedness, flexibility, and adapting to changing circumstances.

This guide provides the structure, but your ability to follow through with the plan will ultimately determine your success. With the right mindset, your home can become a fortress of safety, comfort, and self-sufficiency in the face of any crisis.

Essentials Checklist for Immediate Readiness

When disaster strikes, whether it's a natural disaster, a power grid failure, or any unforeseen emergency, the difference between success and failure often comes down to preparedness. The first step in any emergency plan is ensuring that you have the essential supplies and tools on hand to survive the first crucial hours and days. This Essentials Checklist for Immediate Readiness will guide you through the items every household should have ready to ensure your family's safety, security, and well-being when the unexpected happens.

1. Water Supply

Water is life. In a survival scenario, your ability to access clean, safe drinking water is paramount. Dehydration can begin within hours without water, and it is the most critical resource during any crisis.

Key Considerations:

Amount: Store at least one gallon of water per person per day for drinking and sanitation. For a family of four, a minimum of 12 gallons for three days is recommended.

Water purification: In case your water source becomes contaminated or unavailable, have a means of purifying water. This can include:

- Water purification tablets or drops
- Portable water filters (e.g., LifeStraw, Berkey filters)
- Boiling pots for purification (ensure you have a heat source like a camping stove or grill)
- Water storage containers like clean barrels, jugs, or bottles

2. Food Supplies

During a crisis, access to fresh food may be limited. You must have a stash of non-perishable foods that are easy to prepare and provide adequate nutrition. Stock up on items that require little to no preparation, so you can rely on them when needed.

Essential Food Items:

- Canned goods: Soup, beans, vegetables, and meats (e.g., tuna, chicken, or chili)
- Freeze-dried meals: Lightweight, long-shelf-life meals for camping or survival situations
- Dried fruits and nuts: High-energy, easy-to-store snacks
- Granola bars, protein bars, or energy bars
- Powdered or evaporated milk
- Rice, pasta, or instant noodles
- Peanut butter and other spreads for protein
- Special dietary items: Ensure you include foods for any dietary restrictions (e.g., gluten-free, low-sodium, or vegan items)

Make sure to also account for:

- **Manual can opener:** Essential for opening canned goods in case of a power failure
- **Long-term storage options:** Food vacuum-sealing systems or Mylar bags with oxygen absorbers can extend the shelf life of certain items.

3. First Aid Kit

Accidents and injuries are common during disasters. A well-stocked first aid kit can help address minor injuries and stabilize more serious ones until professional help arrives.

Essential First Aid Supplies:

- **Bandages:** Adhesive bandages, gauze pads, and medical tape
- **Antiseptic wipes and ointments:** Alcohol wipes, iodine, and antibiotic ointments
- **Pain relievers:** Ibuprofen, acetaminophen, aspirin (if appropriate)
- **Tweezers and scissors:** For splinters, glass, and other foreign objects
- **Thermometer:** A reliable, non-contact thermometer
- **Instant cold packs:** To reduce swelling or pain
- Burn gel or cream
- **Antihistamines:** For allergic reactions
- **Prescription medications**: Ensure you have a 72-hour supply of any critical medications for your family, including any vital chronic medications
- **First-aid manual:** A comprehensive guide on how to handle basic medical situations in the absence of professional care

4. Flashlight and Batteries

When the power goes out, a flashlight is essential to navigate through the dark. Stock extra batteries to ensure you can maintain visibility throughout an emergency.

Key Considerations:

- **LED flashlights:** Energy-efficient and long-lasting
- **Headlamps:** Keep your hands free while providing light
- **Solar-powered flashlights:** Great for long-term use without needing batteries
- **Spare batteries:** Make sure you have batteries for all devices (flashlights, radios, etc.)

5. Multitool and Tools

Having a multitool on hand can make all the difference in a survival situation. From opening cans to fixing gear, a reliable multitool is invaluable.

Essential Tools:

- **Multitool:** A versatile tool with knives, pliers, screwdrivers, and scissors
- **Wrench and pliers:** For turning off utilities or fixing broken items

- **Duct tape:** A lifesaver in emergencies for repairs and improvisation
- **Basic set of screwdrivers and nails:** To secure your shelter or equipment
- **Axe or hatchet:** For chopping firewood or cutting through debris
- **Shovel:** Useful for digging, sanitation, or burying waste

6. Shelter and Warmth

Staying warm and dry is crucial during a crisis. The right shelter can keep you safe from the elements, while sleeping bags or blankets will ensure you can rest comfortably.

Shelter Essentials:

- **Tent or tarp:** A shelter for when you're outdoors or if your home becomes uninhabitable
- **Sleeping bags:** Ensure they are rated for your climate
- **Emergency space blankets:** Lightweight, compact, and designed to retain body heat
- **Thermal clothing:** Layering is key for warmth; include hats, gloves, socks, and thermal undergarments
- **Heaters:** Battery-powered or camping heaters (if safe to use indoors)

7. Fire Starting Supplies

Having the ability to start a fire is a critical skill in any emergency situation. It will help you boil water, cook food, keep warm, and signal for help.

Fire Essentials:

- **Waterproof matches or lighters:** A reliable and easy way to start a fire
- **Firestarter sticks or fatwood:** To quickly catch fire in any condition
- **Tinder:** Dry kindling, cotton balls with petroleum jelly, or magnesium fire starters
- **Portable stove:** A small, camp-style stove or a propane burner if cooking becomes necessary

8. Communication Tools

In the chaos of a disaster, it's essential to stay connected with others for updates, resources, and safety information. Whether it's reaching loved ones or receiving emergency broadcasts, reliable communication is vital.

Communication Essentials:

- **Battery-powered or hand-crank radio:** This will allow you to tune into emergency broadcasts without relying on electricity
- **Mobile phone:** Keep it charged with a solar charger or power bank
- **Whistle:** An effective tool for signaling for help, especially when you're outdoors

9. Personal Hygiene and Sanitation

Maintaining hygiene during an emergency is crucial to avoid illness and stay healthy. A basic hygiene kit can prevent the spread of diseases and help everyone feel more comfortable.

Sanitation Essentials:

- **Hand sanitizer:** With alcohol content of at least 60% to kill germs
- **Toiletries:** Toilet paper, wet wipes, feminine hygiene products, and a portable toilet if necessary
- **Trash bags:** To store waste and keep your area clean
- **Face masks:** To protect against dust, smoke, or contamination from biohazards
- **Disinfectant:** To clean surfaces and prevent the spread of germs

10. Important Documents and Cash

When infrastructure collapses, digital communication systems may fail, and banking systems could be down. Keep physical copies of important documents and some cash on hand.

Important Documents:

- IDs and passports
- **Medical records:** Prescription lists, allergy information, immunization records
- **Insurance papers:** Home, health, and life insurance documents
- **Emergency contacts:** A list of phone numbers and addresses for family members, neighbors, and friends

Cash:

Small bills: Cash can become the only currency in a crisis. Have small denominations of bills (ones, fives, tens) in case ATMs and banks are down.

11. Miscellaneous Essentials

Finally, there are a few miscellaneous but critical items you shouldn't overlook. These will help with communication, comfort, and addressing unexpected situations.

Miscellaneous Essentials:

- **Maps:** Printed maps of your local area and surrounding regions in case you need to evacuate
- **Pen and paper:** For note-taking or leaving messages
- **Backup chargers:** Solar-powered or hand-crank chargers for your electronics
- **Entertainment items:** Books, board games, or card games to keep morale up during stressful times

Being prepared is the first step to surviving an emergency. This Essentials Checklist for Immediate Readiness includes the must-have items to ensure your family's survival in the first crucial hours and days of a disaster. Stocking up on these supplies and organizing them ahead of time will give you peace of mind and ensure that when disaster strikes, you are ready.

Common Mistakes to Avoid

When preparing for emergencies, it's easy to get overwhelmed by the sheer volume of information and the urgency of the situation. Many people make mistakes in the process of bugging in—whether they are just beginning their preparedness journey or are seasoned preppers. Avoiding these common mistakes can mean the difference between success and failure when an emergency strikes. Here's a deep dive into the most common missteps preppers make when preparing to bug in, and how to avoid them.

1. Not Planning for Long-Term Needs

One of the most common mistakes preppers make when planning to bug in is focusing only on short-term survival, like getting through the first 72 hours. While the first few days are critical, it's essential to think beyond that. Emergencies can last much longer than three days, and you need to plan for long-term survival.

Mistake to Avoid:

Focusing solely on short-term supplies such as water, food, and first aid kits.

What to Do:

Make sure you have a plan for sustainability in the long term. Stockpile enough supplies to last weeks or even months, and focus on:

- Water filtration systems to purify water when your supply runs out.
- Seeds and gardening tools for growing your own food.
- Preserved food such as freeze-dried meals, canned goods, and MREs (Meals Ready-to-Eat) for long-term consumption.
- Alternate heating sources like wood-burning stoves or propane heaters if electricity is unavailable.
- Power generation such as solar panels, wind turbines, or generators to keep essential systems running.

By thinking about the long haul, you can avoid panic when your initial food and water supplies are depleted.

2. Ignoring Proper Food Storage

Many preppers make the mistake of stocking up on food without properly considering the shelf life, storage conditions, and rotation methods. Food that is improperly stored may spoil or lose its nutritional value over time, rendering it useless when you need it most.

Mistake to Avoid:

Buying large quantities of food without proper storage.

What to Do:

- Store food in cool, dry places to maximize its shelf life. Ideally, store food in airtight containers or Mylar bags with oxygen absorbers.
- Rotate your stockpile regularly. Use the FIFO (First In, First Out) method, meaning you use your oldest items first and replace them with new ones.
- Pay attention to expiration dates. Some foods have a longer shelf life than others, so plan accordingly. Freeze-dried and dehydrated foods can last for several years, while canned goods last 1-5 years depending on the type.
- Include food items that don't require cooking in case you lose the ability to heat food, such as energy bars, dried fruits, and canned soups.

Food storage is one of the easiest aspects of preparedness to overlook, but proper storage can make your food supplies last much longer.

3. Underestimating the Importance of Water

Water is the most essential resource for survival, yet many people make the mistake of not storing enough or failing to plan for water purification. In a long-term crisis, you can survive without food for weeks, but only a few days without water.

Mistake to Avoid:

Storing an insufficient amount of water or failing to account for the need to purify water.

What to Do:

- Store at least one gallon of water per person per day for drinking and sanitation. For a family of four, store at least 12 gallons for a three-day period.
- Invest in water purification systems. Have water filters (such as LifeStraw or Berkey filters), purification tablets, and even a portable solar still in case you need to purify local water sources.
- Store water in multiple containers such as large barrels, jugs, or water bricks to maximize your storage capacity and flexibility.
- Consider a rainwater collection system to supplement your stored water. Ensure it's safe to use by filtering the water before consumption.

Properly managing water can be the difference between life and death in an emergency situation, so prioritize it early in your preparedness plan.

4. Failing to Consider Security

While the goal of bugging in is often to shelter in place, you must also consider the potential for external threats. In a crisis, security can become an issue as desperate individuals or groups may seek to take advantage of vulnerable homes.

Mistake to Avoid:

Not having a comprehensive security plan for your home.

What to Do:

- Fortify your home. Make sure windows and doors are reinforced, and have the necessary tools (wood, nails, locks, etc.) to secure them if needed.
- Set up a perimeter. Knowing the layout of your property and securing vulnerable access points is key. Consider investing in motion sensor lights, cameras, or even temporary fencing if necessary.
- Have self-defense measures in place. Whether you use firearms (if legal in your area), pepper spray, or other personal defense tools, ensure everyone in your household is trained and prepared to defend themselves if the situation escalates.
- Create a communication plan. In case you need help, having a plan for how to contact emergency services or trusted neighbors is vital.

While security might seem like an afterthought, it can make all the difference when others are also scrambling for resources.

5. Not Involving the Whole Family

Another common mistake is preparing without involving all members of the household in the planning process. If only one person knows what to do in an emergency, it can create chaos when that person is unavailable or incapacitated. Furthermore, the psychological aspects of survival should not be overlooked.

Mistake to Avoid:

Not training your family or household on what to do during an emergency.

What to Do:

Involve everyone. Create a family emergency plan that details what each person's role will be. This should include how to find food and water, how to communicate with each other, and what the emergency signal or code will be.

Practice drills regularly so that everyone is comfortable with the plan. Practice evacuation routes, sheltering in place, and how to use emergency supplies.

Teach basic survival skills. Everyone in your household should know how to:

- Use a first aid kit
- Start a fire

- Purify water
- Use a map and compass
- Defend the home if necessary

Training the entire family ensures that no one is left behind, and the household can function cohesively during a crisis.

6. Overcomplicating Things

Preppers often make the mistake of overcomplicating their bug-in plans. It's easy to get caught up in trying to have the perfect survival scenario, but this can lead to unnecessary spending, confusion, and failure to act.

Mistake to Avoid:

Overloading on unnecessary gadgets or complex plans that are hard to implement in the moment.

What to Do:

- Simplify your plan. Focus on the essentials: water, food, shelter, security, and basic medical needs. While advanced gadgets may seem appealing, having too many tools that you're unfamiliar with can actually slow you down during an emergency.
- Prioritize practical solutions over novelty items. While tactical gear and survival tools have their place, know that practical items such as a basic first aid kit, water filters, and canned food are often more crucial than expensive, high-tech survival gear.

A simple, well-organized plan is often far more effective than an over-complicated one that can be difficult to implement when you're under stress.

7. Not Staying Updated

Lastly, a common mistake preppers make is failing to stay up to date with their preparedness plans. Preparedness isn't a one-time activity; it's an ongoing process. New threats, better technology, and new resources become available all the time, so it's critical to stay informed.

Mistake to Avoid:

Letting your preparedness plan go stale.

What to Do:

- Review your supplies regularly. Make sure everything is in working order, especially things like batteries, fire-starting materials, and food supplies.
- Keep informed. Follow news about global events, local disasters, and new survival techniques to adapt your plan accordingly.
- Adapt to changing circumstances. If your family situation changes (e.g., a new child, aging parent, or new pet), update your plan to reflect these changes.

3. UNDERSTANDING THREATS AND RISKS

Identifying Local and Global Emergencies

Being prepared for emergencies is not only about having supplies and survival tools on hand; it's also about understanding the different types of emergencies that can occur, both locally and globally. The ability to identify potential threats is key to effectively preparing for them. This knowledge allows you to tailor your survival strategy and ensure that you are ready for a wide range of scenarios, whether they are small-scale, localized events or larger, global crises.

In this section, we will explore how to identify different types of emergencies, both local and global, and how understanding these threats can help you be better prepared.

What Constitutes a Local Emergency?

A local emergency is an event or situation that primarily affects your immediate area, such as your neighborhood, town, city, or region. These emergencies can range from natural disasters to human-made events and can vary significantly in scale and duration. Some of the most common types of local emergencies include:

1. Natural Disasters

Local natural disasters are some of the most common types of emergencies you might face. They include:

- **Earthquakes:** While earthquakes are common in certain areas, they can occur unexpectedly. A major earthquake can disrupt infrastructure, cause structural damage to buildings, and result in power outages.
- **Flooding:** Local flooding can occur after heavy rainfall or snowmelt, especially in low-lying areas or near rivers. Flooding can damage homes, displace people, and contaminate water supplies.
- **Wildfires:** In areas prone to dry conditions, wildfires can quickly spread, threatening homes, businesses, and even lives. Wildfires are often exacerbated by high winds and extreme temperatures.
- **Tornadoes:** Tornadoes can occur with little warning and cause devastating damage to homes, schools, and entire communities, particularly in tornado-prone areas like the Midwest and South of the U.S.
- **Severe Weather Events:** Extreme weather events like hurricanes, blizzards, heatwaves, and hailstorms can lead to power outages, property damage, and other emergencies.

2. Human-Made Disasters

While natural disasters often receive the most attention, human-made disasters can also cause widespread disruption in local areas. These include:

- **Industrial Accidents:** Chemical spills, explosions, or nuclear accidents can have local environmental impacts, affecting air quality, water safety, and food supplies.
- **Mass Casualty Incidents:** Events such as bombings, shootings, or terrorist attacks can have immediate and devastating consequences for communities, leading to both short-term and long-term emergencies.
- **Infrastructure Failures:** Power grid failures, water supply contamination, and transportation system breakdowns can create significant disruptions, leaving communities vulnerable to further complications.
- **Pandemics:** While some pandemics have global reach, their effects can start as localized outbreaks before they spread. The COVID-19 pandemic is an example of how a local health emergency can rapidly escalate into a global crisis.

What Constitutes a Global Emergency?

A global emergency is an event or crisis that affects large swaths of the population, often extending beyond national borders. These emergencies can be far-reaching, both in terms of their geographic impact and their ability to disrupt multiple aspects of daily life, such as economies, politics, and the environment. Some of the most notable global emergencies include:

1. Global Pandemics

The COVID-19 pandemic is a recent and stark reminder of how quickly a localized disease outbreak can escalate into a global crisis. Pandemics can cause widespread illness, overwhelm healthcare systems, disrupt the global economy, and create social and political upheaval. Pandemics can spread through air travel and other forms of global transportation, making them difficult to contain once they begin spreading rapidly.

Key considerations for pandemics:

- The ability of healthcare systems to cope with large numbers of sick people.
- The availability and distribution of vaccines and medical supplies.
- The need for global cooperation in containing and responding to the spread of disease.

2. Climate Change and Environmental Disasters

Global climate change is an ongoing, slow-moving crisis that is creating more frequent and severe weather events. As the global climate shifts, extreme weather patterns are becoming more common, and environmental disasters are impacting entire regions and, in some cases, the entire planet. These include:

- **Rising Sea Levels:** Coastal cities worldwide are facing the risk of flooding and displacement due to rising sea levels. Cities like Miami, New York, and Tokyo are experiencing this firsthand.
- **Severe Weather Events:** Heatwaves, droughts, floods, and hurricanes are intensifying and becoming more frequent due to climate change. The devastation caused by global hurricanes like Hurricane Katrina and Typhoon Haiyan illustrate the magnitude of these events.
- **Loss of Biodiversity:** Environmental degradation, including deforestation, soil erosion, and the depletion of natural resources, can lead to the collapse of ecosystems, which in turn disrupt food systems and global economies.

3. Geopolitical Crises

International conflicts, wars, and political instability are increasingly a global concern. Political tensions can escalate quickly into conflicts that spread across borders and affect neighboring countries, creating refugees and economic instability. For example:

- **Armed Conflicts and Wars:** Wars can lead to refugee crises, economic collapse, and widespread famine. The ongoing conflicts in Syria, Ukraine, and Afghanistan highlight how war and instability can affect millions of people globally.
- **Nuclear Threats:** The threat of nuclear conflict, whether through warfare or the spread of nuclear weapons, remains one of the most significant global risks. The tensions between nuclear powers like the U.S., Russia, and North Korea have the potential to spark global emergencies.

4. Global Economic Collapse

Economic instability and the potential for a global economic collapse are critical risks that can impact everyone, from individuals to entire nations. Financial crises, such as the 2008 global recession, can lead to massive unemployment, a loss of savings, and an overall reduction in the quality of life. In extreme cases, these financial crises can lead to hyperinflation, civil unrest, and social breakdowns.

Identifying and Preparing for Both Local and Global Emergencies

To be fully prepared, it's important to consider both local and global emergencies. While it's impossible to predict every crisis, you can identify potential threats based on your location and circumstances, and then prepare accordingly. Here are some steps to help you identify and prepare for these emergencies:

1. Conduct a Risk Assessment

- **Local Threats:** Research the most common disasters in your area. Look into historical data to understand which natural and human-made emergencies are most likely to occur. For example, if you live in California, you might prioritize earthquake preparedness, while those in Florida might focus on hurricanes.

- **Global Threats:** Stay informed about global events that could affect you, such as pandemics, economic instability, or geopolitical tensions. Subscribe to reliable news sources and global watchlists that track crises and their potential impact.

2. Plan for Local Emergencies

- **Tailor Your Preparedness Plan:** Once you identify potential local threats, customize your emergency plan and kit to address those specific risks. This might mean having flood barriers in place, having an evacuation route mapped out, or maintaining a large water supply for drought conditions.
- **Focus on Self-Sufficiency:** For local emergencies, your goal is often self-sufficiency. Ensure you have sufficient resources for at least 72 hours, with redundancy for essentials like food, water, and medical supplies.

3. Plan for Global Emergencies

- **Stay Informed and Flexible:** Global emergencies are often more unpredictable and widespread, so you need to be ready for more diverse and complex scenarios. Stay informed through news outlets, international organizations, and governments' advisory services.
- **Focus on Long-Term Preparedness:** When preparing for global emergencies like pandemics or climate change, focus on longer-term sustainability. Consider food storage, off-grid power solutions, and community-building as part of your strategy.

Identifying local and global emergencies is the first step in crafting a comprehensive survival plan. By recognizing the potential risks in your area and globally, you can begin to prepare effectively for a range of scenarios. Whether it's a localized earthquake or a global pandemic, being prepared is not just about survival—it's about resilience. By understanding the threats that could impact you, you can design a plan that gives you the best chance to survive and thrive, no matter the crisis.

Home Vulnerability Assessment

A home vulnerability assessment is a critical step in preparing for emergencies, whether they are local or global in nature. It involves evaluating your home, property, and surroundings to identify potential risks and weaknesses that could impact your safety and security during a disaster or survival scenario. By understanding where your home is vulnerable, you can take proactive measures to fortify it and increase your chances of survival.

In this section, we will look deep into how to conduct a comprehensive home vulnerability assessment, covering everything from structural integrity to security systems, emergency preparedness, and environmental hazards.

Why Conduct a Home Vulnerability Assessment?

The importance of conducting a vulnerability assessment lies in its ability to help you identify weak spots and address them before an emergency occurs. Disasters can strike at any time, and being caught off-guard can lead to unnecessary damage, injury, or loss of life. A well-prepared home, on the other hand, provides a safe refuge that can mitigate the risks posed by external threats.

Here are some reasons why a vulnerability assessment is essential:

- **Increase Safety:** Identifying hazards and potential risks allows you to make necessary repairs and upgrades that will enhance your home's ability to withstand an emergency.
- **Improved Emergency Planning**: A vulnerability assessment helps you develop a tailored emergency plan for your household, ensuring that everyone knows what to do and where to go during an emergency.
- **Peace of Mind:** By proactively addressing weaknesses in your home, you gain confidence that you are better prepared for whatever may come.

Key Areas to Include in a Home Vulnerability Assessment

A thorough home vulnerability assessment covers several key areas, ranging from the structural integrity of your home to the security measures you have in place. Below are the primary components to focus on during the assessment.

1. Structural Integrity

The foundation of your home is one of the most crucial elements in determining how well it will fare during an emergency. Natural disasters like earthquakes, floods, or hurricanes can expose weaknesses in the structure, potentially causing significant damage. Here are key factors to evaluate:

- **Foundation:** Inspect the foundation for cracks, shifts, or signs of weakness. In areas prone to flooding, make sure that the foundation is elevated above potential water levels. For earthquake-prone regions, ensure that the foundation is reinforced to withstand seismic activity.
- **Roof:** Check the roof for missing shingles, leaks, or signs of wear. In hurricane-prone regions, consider installing impact-resistant roofing materials to withstand high winds and flying debris.

- **Windows and Doors:** Reinforce windows and doors to prevent breakage during storms or intrusions. Consider installing storm shutters or impact-resistant glass. Ensure that all entry points are properly sealed and secure.
- **Walls and Structure:** Examine the exterior and interior walls for any cracks or signs of deterioration. For areas prone to earthquakes or strong winds, you may need to add structural supports or bracing.

2. Security Measures

Your home's security is another essential area to assess, as a well-secured home reduces the risk of intruders or looters during an emergency. Here's what to consider:

- **Locks and Deadbolts:** Ensure all external doors are equipped with high-quality locks and deadbolts. Check that locks are functioning properly and secure.
- **Fencing and Perimeter Security:** Evaluate the strength and height of fences or walls surrounding your property. Secure gates and consider adding locks or security systems to deter unauthorized access.
- **Security Systems:** Install security cameras, motion detectors, and alarm systems to monitor your property. Many modern security systems can be linked to your phone for remote monitoring, which is crucial in emergency situations.
- **Lighting:** Install exterior lighting to illuminate pathways, driveways, and entry points. Motion-activated lights are particularly effective at deterring intruders.

3. Emergency Power and Water Supply

During an emergency, power and water are two of the most critical resources. Without access to electricity or clean water, you may find yourself in a dire situation. Here's how to assess your preparedness in these areas:

- **Backup Power:** Invest in a backup generator to supply power in case of a grid failure. Ensure that the generator is appropriately sized for your needs and that it is located safely away from windows and doors to prevent carbon monoxide buildup.
- **Water Storage:** Store enough clean water for your family for at least three days, with one gallon per person per day. Consider installing a rainwater harvesting system or storing water in large containers or barrels.
- **Water Filtration:** Have a reliable water filtration system in place, especially if your water supply is prone to contamination. You can purchase portable water filters or invest in a larger home filtration system to purify rainwater, well water, or emergency supplies.

4. Fire Safety

Fire is one of the most common hazards during emergencies, particularly in areas prone to wildfires, floods, or industrial accidents. A fire can easily spread through a home if proper precautions aren't in place. To assess fire safety in your home:

- **Smoke Detectors:** Ensure that smoke detectors are installed in all major rooms, especially near sleeping areas. Test them regularly to ensure they're functioning and replace the batteries as needed.
- **Fire Extinguishers:** Keep fire extinguishers in key areas of your home, such as the kitchen, garage, and near any heating or electrical appliances. Make sure everyone in your household knows how to use them.
- **Escape Plan:** Develop and practice a fire evacuation plan with your family. Identify multiple exit points from each room, and designate a safe meeting place outside the home. Make sure all family members, including children, understand the plan.

5. Environmental Hazards

Certain environmental hazards, such as floods, landslides, and radon, can pose a significant risk to your home. In order to mitigate these risks, take the following steps:

- **Flooding Risks:** If you live in an area prone to flooding, consider elevating your home's foundation or installing sump pumps. Ensure that your home is equipped with waterproof barriers and that gutters are regularly cleaned to prevent blockages.
- **Radon and Toxic Gases:** Radon is a naturally occurring gas that can infiltrate your home and cause serious health issues. Test your home for radon levels and take corrective action if necessary. Also, consider installing a carbon monoxide detector to monitor for dangerous levels of this odorless gas.
- **Pests and Rodents:** Pests, including termites, rodents, and insects, can weaken your home's structure and create unsanitary living conditions. Inspect your home for signs of infestation and take steps to eliminate any pests before they become a serious problem.

6. Communication and Signaling

During an emergency, communication is essential. It's crucial to have a system in place to stay informed about developments and to alert others if needed. Here's how to assess your communication preparedness:

- **Emergency Radio:** Invest in a battery-powered or hand-crank emergency radio to receive weather alerts, emergency broadcasts, and updates from local authorities, especially if power is lost and cell networks become overloaded.
- **Cell Phones and Backup Power:** Ensure that your cell phones are fully charged and that you have backup power sources like portable chargers or solar-powered chargers to keep your devices operational during an extended outage.

- **Emergency Signaling Devices:** Have visual signaling devices, such as emergency flares or bright-colored flags, in case you need to signal for help. You can also use reflective tape around windows or doors to alert rescue teams to your location.

7. First Aid and Medical Preparedness

Health and safety are paramount in any emergency situation. To prepare your home, make sure you have:

- **First Aid Kits:** A well-stocked first aid kit is essential for handling injuries, cuts, burns, and other medical emergencies.
- **Prescription Medications:** Keep an extra supply of any prescription medications you or family members rely on. Consider storing these in a cool, dry place and rotating them regularly to prevent expiration.
- **Medical Supplies:** Depending on your family's needs, you may need to store additional medical supplies, such as bandages, antiseptics, splints, and medical tools for more severe emergencies.

A home vulnerability assessment is the first step in safeguarding your household from a range of potential emergencies. By taking the time to evaluate your home's weaknesses and making the necessary improvements, you can significantly increase your family's safety and readiness during an emergency. From structural integrity and security measures to emergency power and fire safety, a thorough assessment will ensure that your home is prepared for the challenges that lie ahead. Remember, proactive preparation is key to survival, and a secure, well-prepared home is the best foundation for weathering any storm—natural or man-made.

Deciding Between Bugging In and Bugging Out

When disaster strikes, one of the most crucial decisions you will face as a prepper is whether to bug in or bug out. Both strategies are designed to ensure your survival during an emergency, but they require vastly different approaches, resources, and mindsets. The choice between staying in your home (bugging in) and fleeing to a safer location (bugging out) depends on various factors, including the nature of the threat, your current situation, and your preparedness level. Understanding the pros and cons of each option, as well as the key factors influencing your decision, can help you make the right call when the time comes.

In this chapter, we will explore both bugging in and bugging out in depth, providing the information and tools you need to decide which strategy is best suited for you and your family in different situations.

What Does It Mean to Bug In?

Bugging in refers to staying in your home or a safe, fortified location during an emergency. The primary goal of bugging in is to secure your position and wait out the crisis. This option is often considered the best choice for urban dwellers or those who live in areas with higher risks of civil unrest, wildfires, or severe weather events. Bugging in allows you to utilize your home's resources, such as food, water, and medical supplies, while maintaining access to security systems and shelter.

Advantages of Bugging In

Familiarity with Your Environment: Your home is familiar, which means you know the layout, the resources available, and any potential vulnerabilities. This gives you an advantage in a crisis, as you can quickly adapt to changes and respond more efficiently.

Access to Supplies and Resources: When bugging in, you have immediate access to your stockpile of food, water, first aid supplies, and any other preparedness materials you have stored. There's no need to worry about rationing your supplies while trying to transport them to a new location.

Security: If your home is well-prepared and fortified, bugging in can offer significant security benefits. You have control over who enters your space, and with the right security systems in place, you can monitor any external threats and take steps to protect your home.

Less Risk of Exposure: Bugging in reduces your exposure to the dangers that might be present outside, such as looters, contaminated air, or extreme weather conditions. You also avoid the risks associated with traveling, like encountering roadblocks, hostile forces, or getting lost.

Comfort: Staying in your home can be much more comfortable than bugging out, particularly if you have established a safe and well-stocked emergency refuge. You have a familiar environment, your personal belongings, and family members close by, which can reduce stress and increase your morale during difficult times.

Disadvantages of Bugging In

- **Limited Mobility:** One of the main drawbacks of bugging in is that you are confined to a fixed location. If the situation worsens or changes unexpectedly, such as a wildfire approaching or civil unrest spreading, you may find yourself trapped with no means of escape.
- **Vulnerability to Siege Situations:** If you are not well-prepared, bugging in can leave you vulnerable to a siege situation where your home could be surrounded, cut off from essential resources, or overrun by hostile forces. The longer you remain inside, the more difficult it may be to maintain resources, manage food, and defend your position.
- **Infrastructure Breakdown:** Depending on the disaster, essential services like electricity, water, and gas may be interrupted, and you could face long-term shortages. In this case, even if you stay inside, you'll need alternative sources for power, water, and food to survive. This requires significant planning and preparedness.

What Does It Mean to Bug Out?

Bugging out, on the other hand, involves leaving your home and relocating to a safer area in response to an imminent threat. This is usually done when staying at home becomes too dangerous, whether because of natural disasters, civil unrest, or military conflict. Bugging out may involve heading to a designated bug-out location, which could be a rural retreat, a family member's home in a safer area, or a pre-established shelter in the wilderness.

Advantages of Bugging Out

- **Escape from Immediate Danger:** If your home is in a direct path of danger, such as a wildfire, flood, or nuclear fallout, bugging out can be the best way to avoid harm. It allows you to move away from the threat and find a safer, more secure environment.
- **Access to Safer Territory:** Bugging out allows you to move to an area where resources are more abundant, or where the infrastructure is more stable. This might include relocating to a rural location where there are fewer people, lower risks of disease, and more space to grow food or shelter.
- **More Control Over Your Survival:** By leaving your home, you have more control over your environment. You can choose a more suitable location, access more land, and be away from potentially hostile situations, such as urban areas with high crime rates.
- **Reduced Risk of Overcrowding:** In the event of a large-scale disaster, urban areas can quickly become overcrowded, with looting, violence, and other threats becoming more pronounced. Bugging out to a more remote area can help you avoid these issues and give you a better chance of survival in peace.

Disadvantages of Bugging Out

- **Uncertainty and Risk of Getting Lost:** Bugging out involves navigating unknown territory, whether that's rural land or unfamiliar urban areas. There's always a risk of getting lost, encountering hostile individuals or groups, or finding that your designated bug-out location is compromised or unreachable.

- **Transportation Challenges:** Depending on the nature of the disaster, roads could be impassable, or fuel shortages could make it difficult to travel. Without reliable transportation, you may find yourself stranded or stuck in a location far from your intended bug-out destination.
- **Limited Resources:** While a bug-out location may offer more safety, it may also lack the resources and amenities you've come to rely on at home. You might find yourself with limited access to food, water, medical supplies, or shelter until you can establish a new base of operations.
- **Loss of Familiarity:** Unlike bugging in, bugging out requires you to leave the comfort of your home and your familiar surroundings. This can create a lot of psychological stress, especially if you're traveling with children, pets, or elderly family members. There's also the emotional toll of leaving behind your personal possessions and the memories tied to them.

Factors to Consider When Deciding Between Bugging In and Bugging Out

Choosing between bugging in and bugging out isn't always clear-cut. Several factors influence the decision, and it's important to be prepared for both scenarios. Here are some of the key considerations to keep in mind:

1. Type of Disaster

- **Natural Disasters:** Events like hurricanes, tornadoes, or wildfires may dictate the need to bug out, especially if your area is directly impacted. On the other hand, if you live in an area prone to flooding or earthquakes, bugging in may be a more appropriate choice, provided your home is adequately fortified.
- **Civil Unrest or Societal Collapse:** In cases of widespread civil unrest, violent protests, or economic collapse, bugging in can provide you with a more secure position, particularly if you have a well-stocked home with adequate security systems. However, if the unrest becomes widespread and your home is at risk of being looted, bugging out may be the safer option.
- **Pandemics or Disease Outbreaks:** In the case of a health emergency, bugging in can help you avoid exposure to a contagious disease. With quarantine measures in place, leaving your home could increase your chances of infection, while staying put allows you to isolate yourself.

2. Available Resources

- **Food and Water Supply:** Bugging in is the best option if you have sufficient food, water, and medical supplies at home. If your resources are limited or you need to access a better supply, bugging out might become necessary.
- **Security:** If your home is well-fortified and you have security systems in place, bugging in might provide the best safety. However, if your security is weak or if you're at risk of being overrun, bugging out may be the only viable option.

3. Duration of the Crisis

- **Short-Term Emergencies:** In cases of temporary disruptions, such as a short-term power outage or a localized natural disaster, bugging in is usually the best option.

- **Long-Term Disasters:** For long-term, ongoing crises, especially those that cause a breakdown of law and order, bugging out to a remote location may provide more long-term survival benefits.

4. Family Considerations

- **Children and Elderly:** If you have young children or elderly family members, the decision between bugging in and bugging out may be influenced by their mobility, medical needs, and comfort. A bug-in situation can be less stressful for these groups as it provides them with familiar surroundings.
- **Pets:** If you have pets, consider their needs in both scenarios. Bugging out may be difficult if you need to transport animals, but staying home can ensure their safety as long as you have the resources to care for them.

The decision between bugging in and bugging out is complex and depends on a wide range of factors. Both strategies have their advantages and disadvantages, and each scenario requires careful planning and preparation. By assessing the nature of the threat, the available resources, and your family's needs, you can make an informed decision that maximizes your chances of survival. Ultimately, the key to success lies in being flexible, adaptable, and well-prepared for any situation that may arise.

4. CREATING A FAMILY-CENTERED BUG-IN PLAN

Building a Custom Emergency Blueprint

Creating a personalized emergency blueprint is one of the most effective strategies for ensuring that you and your family are well-prepared for any disaster. Whether it's a natural disaster, a man-made crisis, or a societal breakdown, an emergency blueprint serves as a comprehensive plan that outlines your response to a wide range of potential emergencies. By building a custom emergency blueprint, you tailor your preparedness plan to your specific needs, resources, and location. This guide will walk you through the essential steps to craft an emergency blueprint that suits your lifestyle and helps ensure the safety and well-being of those you care about.

Step 1: Assess Your Risks and Threats

The first step in creating your custom emergency blueprint is identifying the potential threats and risks specific to your area. Understanding the hazards you may face is crucial for developing a tailored plan. The risks you face will vary depending on where you live, your environment, and the current global and local conditions. Some common risks include:

Natural Disasters:

- **Earthquakes:** If you live in a seismic zone, you'll need to plan for earthquakes by securing heavy furniture, knowing safe areas, and preparing for potential aftershocks.
- **Floods:** In flood-prone areas, create an evacuation route and ensure your home is elevated or waterproofed as much as possible.
- **Hurricanes, Tornadoes, and Wildfires:** Coastal or tornado-prone areas require plans for evacuation and fortifying your home.
- **Winter Storms and Extreme Cold:** If you're in a colder region, you must prepare for power outages, heating alternatives, and staying warm without electricity.

Man-Made Disasters:

- **Civil Unrest:** In cities with a history of protests or social instability, knowing where to find safe places or routes out of the city is crucial.
- **Pandemics and Disease Outbreaks:** Understanding how to isolate your family, manage medical needs, and access necessary supplies will be essential during health emergencies.
- **Technological Failures:** Power grid failures, water contamination, or telecommunications breakdowns need to be factored in as well, as they can affect your ability to communicate and access vital services.

Global Threats:

- **Economic Collapse:** Financial crises or hyperinflation can destabilize your community and make it harder to access everyday goods.

- **Nuclear War or Terrorism:** In extreme cases, national security threats may require even more specialized plans for sheltering, evacuation, or survival.

By assessing the primary risks specific to your home and region, you can begin to form the foundation of your emergency blueprint, ensuring that you cover the most likely scenarios.

Step 2: Define Emergency Roles and Responsibilities

A crucial element of any emergency plan is establishing clear roles and responsibilities for each member of your household. This ensures that everyone knows exactly what to do during a crisis, which minimizes confusion and increases your chances of surviving the emergency. Here's how to divide and assign roles:

- **Primary Responders:** Identify the individuals who will take immediate action. This may include tasks such as turning off utilities, securing doors and windows, or administering first aid.
- **Evacuation Coordinators:** Designate someone to ensure everyone in the household is ready to leave the home if necessary, and determine where everyone will meet in case of evacuation.
- **Communicators:** Assign one or two people to act as primary contacts for extended family or neighbors. This person can relay information, update everyone on the situation, and ensure there's a designated way to stay in touch.
- **Supplies Managers:** This role involves overseeing the stockpile of emergency supplies, including food, water, medical kits, flashlights, and other survival gear. They'll also monitor expiration dates and ensure that everything is ready for use when needed.
- **Defenders/Protectors:** In situations involving civil unrest or potential threats from others, designate individuals in charge of defending the home or family, using appropriate tools or strategies.

Make sure everyone in the household understands their roles and practices them through drills or mock emergencies. This ensures that your team is ready to act efficiently when it matters most.

Step 3: Create a Communication Plan

One of the most important components of your emergency blueprint is a communication plan. In many disasters, phone lines or the internet may go down, so it's essential to have alternative communication strategies in place. Your communication plan should include:

1. Contact Information:

- **Family Contacts:** List the phone numbers and addresses of all family members, neighbors, and close friends. Make sure everyone has a hard copy of this list, as well as digital copies stored securely.
- **Emergency Services:** Identify emergency contact numbers, including local authorities, hospitals, fire departments, and shelters. These should be included in your communication plan.

2. Meeting Points:

Primary Meeting Location: Choose a safe place where your family will meet in case of evacuation. This could be a relative's house, a designated public area, or a community center. Make sure everyone knows how to get there.

Secondary Meeting Location: If your primary meeting place becomes unsafe, select a backup location.

3. Out-of-Area Contact:

If local communications are unavailable, designate an out-of-area contact (someone far from your location, perhaps in another state) whom everyone can check in with. This person can serve as a central point for relaying information.

4. Emergency Codes:

Establish emergency codes for different situations. For instance, a specific phrase or word can be used to communicate that you need help or are in danger, without the need for verbal explanations.

Step 4: Develop an Evacuation Plan

No emergency blueprint is complete without a detailed evacuation plan. While bugging in may be the best option in many situations, some emergencies will force you to leave your home. A well-thought-out evacuation plan will help ensure your family's safety and minimize confusion.

1. Identify Routes:

Know the primary routes to take out of your city or town. Have at least two routes in case one is blocked or impassable.

Consider different types of evacuation methods (car, foot, bike) based on what's available to you and the nature of the emergency.

2. Transportation:

Ensure your vehicles are always maintained and fully fueled. Keep spare fuel, vehicle essentials, and emergency tools in your car.

For those without access to a vehicle, plan how to get out using public transportation or by walking. Have backup transportation options available if possible.

3. Know Your Evacuation Destinations:

Have pre-arranged destinations for when you leave. This could be the home of a family member or a designated shelter. Ensure these places are stocked with necessities and are safe to travel to.

4. Animal Evacuation:

Make arrangements for pets or livestock in your evacuation plan. If possible, include a portable pet carrier or other means of transportation for animals.

Step 5: Secure Your Home and Property

Securing your home is a vital part of your emergency blueprint. A home that is safe and well-prepared can offer shelter, comfort, and protection during a disaster. Steps to consider include:

1. Fortifying Your Home:

- **Doors and Windows**: Install reinforced doors and windows with high-quality locks. Consider adding security bars or mesh for windows.
- **Safe Rooms:** If possible, designate a room or space where you and your family can shelter safely. This could be a basement or a room with no windows that can be easily fortified.
- **Perimeter Security:** Use outdoor lighting, motion sensors, and security cameras to monitor the exterior of your home. Additionally, if there is civil unrest, consider setting up barriers or even a simple fence.

2. Backup Power:

Install backup power sources such as solar panels, a portable generator, or a battery-powered system. Ensure you have enough fuel or batteries to last through extended power outages.

3. Stockpile Essential Supplies:

Your blueprint should include a detailed list of necessary supplies, such as food, water, medical kits, sanitation products, batteries, tools, and important documents. Be sure to rotate your supplies regularly to keep everything up to date.

Step 6: Regularly Review and Update Your Plan

Your emergency blueprint is a living document that must evolve as your family's needs change, and as new threats and challenges emerge. Make sure to:

Practice Drills: Conduct emergency drills to familiarize your family with the plan and test your readiness.

Update Information: Review your contacts, routes, and plans regularly. Keep your supplies fresh and add new tools as technology improves.

Adapt to Changing Circumstances: Be open to changing your plan as you learn more about your vulnerabilities and available resources. Prepare for worst-case scenarios, but also stay flexible.

Building a custom emergency blueprint requires time, effort, and careful consideration of your unique needs and risks. By following these steps and creating a comprehensive, personalized plan, you can significantly improve your family's resilience in the face of disaster. This proactive approach will give you peace of mind, knowing that you are prepared for the unexpected and capable of handling any emergency that comes your way.

Communication Strategies for Loved Ones

In times of crisis, one of the most critical factors in ensuring safety and survival is effective communication. During natural disasters, pandemics, civil unrest, or any other type of emergency, maintaining communication with your loved ones can make all the difference between staying informed and losing touch with each other. Whether you are bugging in at home or have to evacuate, a solid communication strategy can help keep everyone connected, coordinate efforts, and ensure everyone is accounted for.

This subchapter will guide you through effective communication strategies that you can implement to stay in touch with loved ones during emergencies, the tools and techniques available, and how to structure your communication plan to ensure it works when needed the most.

Step 1: Understand the Importance of Communication During Emergencies

In an emergency, communication becomes the backbone of survival. During crises, people can easily become separated, disoriented, or injured, and without a means of staying in contact, the situation can deteriorate quickly. Effective communication helps by:

- **Providing Real-Time Information:** In a disaster, the information you receive could be crucial to making the right decisions. From evacuation orders to road closures, communication helps everyone stay informed about the current state of affairs.
- **Coordinating Evacuation and Resources:** When multiple family members or friends are involved, coordinating efforts becomes critical. Knowing where to go, how to get there, and what resources to take along is essential. Communication helps align actions.
- **Ensuring Safety:** In an emergency, it's important to quickly check on the well-being of your loved ones. Whether they're inside the house or further away, being able to confirm their safety will reduce panic and help you make informed decisions.
- **Maintaining Calm and Reducing Anxiety:** The psychological aspect of communication should not be overlooked. Staying in touch with loved ones provides reassurance, which helps reduce stress and anxiety during high-pressure situations.

Step 2: Establish a Communication Plan

Before a crisis occurs, you should have a communication plan in place. This plan will help you and your family stay connected when communication systems fail or become overwhelmed. Key aspects of your communication plan include:

1. Emergency Contacts and Communication Tree

Develop a comprehensive list of emergency contacts, which should include family members, close friends, neighbors, and emergency services. Ensure that every family member has this list stored both digitally (on their phones) and physically (in their emergency kits).

- **Primary Contact Person:** This should be someone outside of your immediate area, ideally in another city or state. This person acts as a central point of communication for your family, where everyone can report their status, and relay information about the emergency situation.
- **Family Communication Tree:** A family communication tree is a hierarchical structure of contacts to ensure that each person knows who to contact next. For instance, if one family member cannot be reached, they can contact another person in the tree. This avoids confusion when local phone lines or cell networks become congested.

2. Designated Meeting Locations

If phone lines or digital communication methods become unavailable, establish designated meeting places both locally (near your home) and at a distance (in case evacuation is necessary). These meeting points should be familiar and safe locations, such as the homes of friends or family members, community centers, or local parks. It's crucial that every family member knows how to get to these locations.

- **Primary Meeting Location:** Choose a place that's easily accessible and safe from the disaster. This is where everyone should go immediately after evacuating.
- **Secondary Meeting Location:** A backup meeting location should be established in case the primary location becomes inaccessible or unsafe.

3. Out-of-Area Emergency Contact

Designate an out-of-area contact who can relay information between family members if local communications fail. Often, long-distance phone lines will work even when local communication systems are down. This person can receive updates from each family member and provide them with information about the situation, especially if you cannot reach each other directly.

Step 3: Leverage Communication Tools and Technology

There are various tools and technologies available to aid in communication during emergencies. While some of these tools might depend on specific technologies, others are more general and reliable for most situations.

1. Mobile Phones and Text Messaging

Mobile phones are often the most reliable form of communication during emergencies, but they can become overloaded during high-traffic periods, or networks may fail entirely. Text messaging can often be more reliable than voice calls because texts require less bandwidth.

- **SMS and Messaging Apps:** Encourage your family to use SMS or messaging apps like WhatsApp, Telegram, or Signal to send quick updates. These apps work over cellular networks or Wi-Fi, and can be more reliable than phone calls when the network is congested.
- **Group Messaging:** Set up group chats for your family or other groups so that multiple people can receive updates at once, which saves time and ensures that everyone is on the same page.

2. Ham Radios and CB Radios

Ham radios and CB (Citizens Band) radios are invaluable communication tools in an emergency. These radio systems allow you to broadcast messages without relying on traditional phone networks. They can be a lifeline when other forms of communication are unavailable, particularly in rural or remote areas.

- **Ham Radio:** Requires a license to operate, but it can be used for long-distance communication during emergencies. These radios can connect with other stations over great distances and are typically not affected by network congestion.
- **CB Radio:** These radios can be used without a license for local communication and are a great option if you're stranded in an area without cell coverage.

3. Satellite Phones and Devices

Satellite phones can work anywhere in the world, even when there is no cellular network. They are especially valuable for preppers who live in remote areas or for situations where you may need to communicate during an extended disaster that overwhelms local infrastructure.

- **Satellite Phones:** These phones are ideal for long-range communication in areas where cellular towers are down. They are more expensive than regular phones, but they can be a worthwhile investment if you live in an area prone to isolated disasters.
- **Satellite Messengers:** Devices like Garmin InReach or SPOT are small, portable satellite messengers that allow for two-way text messaging and emergency SOS alerts. These devices can be invaluable in providing communication when you're in the wilderness or during a grid-down situation.

4. Two-Way Radios (Walkie-Talkies)

For local communication, two-way radios can be very effective, especially if you're in a situation where you can't rely on mobile phones. These radios are typically limited in range but can be useful in keeping family members in touch while you're out searching for supplies or navigating an emergency situation.

- **Family Radio Service (FRS):** This is the most common type of two-way radio for short-range communication (usually up to 5 miles). It's suitable for neighborhood or family communication.
- **General Mobile Radio Service (GMRS):** GMRS radios have a longer range (up to 50 miles with the right conditions) but require a license to operate. They're a great option for larger families or groups.

5. Emergency Alert Systems

Sign up for local emergency alert systems that provide real-time notifications about severe weather, road closures, or other urgent events. Many cities and counties offer free alerts via text, email, or phone calls, which will keep you informed about ongoing events in your area.

- **Wireless Emergency Alerts (WEA):** These alerts are broadcast to mobile phones, providing critical information about natural disasters, evacuation orders, and other safety alerts.
- **NOAA Weather Radio:** A battery-powered or hand-crank weather radio can be used to receive National Weather Service (NWS) alerts, which are particularly useful in rural or isolated areas where cell phone signals might not be strong.

Step 4: Implement Backup Communication Plans

In case communication systems break down entirely, it's important to have backup strategies in place:

Prearranged Signals: Establish a series of signals to convey certain messages, such as "I'm okay" or "I need help." This could include specific phrases, a designated light signal (like flashing a flashlight), or even a colored flag or cloth in a window.

Written Notes and Signs: If you're in an area without any technology, consider leaving notes or signs with key information. For example, a note left on your door or a car window can tell people whether you've evacuated or if you're sheltering in place.

Step 5: Conduct Communication Drills

Finally, a communication plan is only as effective as your practice and preparation. Make sure that everyone in your family knows the communication plan and has practiced it. Conduct drills to ensure everyone knows how to use the communication tools and when to implement them. Simulate various emergencies and see how well the plan works under different conditions.

Effective communication is a vital part of emergency preparedness. By understanding the best tools and strategies available, you can ensure that your loved ones remain connected and informed during any crisis. From setting up a communication tree to using satellite phones, the strategies outlined here can help keep you and your family safe and coordinated when it matters most. By preparing for the unexpected, you can reduce stress, avoid confusion, and take the necessary steps to protect yourself and your loved ones in times of crisis.

Caring for Pets, Children, and the Elderly

When preparing for an emergency, it's essential to consider the specific needs of all members of your household, including pets, children, and the elderly. While adults may be able to understand and respond to an emergency situation with a certain degree of self-sufficiency, pets, children, and elderly individuals require special attention, care, and accommodations. These groups are particularly vulnerable during times of stress and chaos, so ensuring their safety and well-being is paramount. In this subchapter, we will discuss how to adequately care for these important members of your family during an emergency and how to integrate their needs into your overall preparedness plan.

Caring for Pets During Emergencies

Pets are beloved members of many families, but they are often overlooked in emergency preparedness plans. In times of disaster, your pet's safety and well-being are just as important as your own. From evacuations to food shortages and medical needs, pets require careful attention.

1. Preparing Your Pets for Emergencies

Before an emergency strikes, it's essential to prepare your pets with a plan that addresses their unique needs. Consider the following:

Pet Emergency Kit: Just as you would have an emergency kit for yourself, create a pet-specific emergency kit. This should include:

- Food and water for at least 72 hours
- Pet medications and a list of any allergies or medical conditions
- First aid supplies, including bandages, tweezers for splinters or ticks, and pet-safe antiseptic
- Copies of vaccination records, microchip information, and other identification
- Leashes, collars, and harnesses for easy handling
- Comfort items, such as a favorite blanket or toy, to help reduce stress
- Carrier or crate for transport, especially for smaller pets

Identification: Ensure your pets have proper identification, such as a collar with an ID tag or, ideally, a microchip. In the event that your pet gets separated, proper identification can make reuniting much easier.

Evacuation Plan for Pets: In case of evacuation, determine ahead of time which shelters or hotels accept pets. Not all emergency shelters allow pets, so plan accordingly. If you must evacuate, ensure that your pet is familiar with its carrier or crate and that you have a safe, accessible means of transportation.

2. Health and Medical Considerations

Pets often have specialized medical needs that must be addressed during an emergency:

Medications: If your pet is on medication, make sure you have enough supplies to last for at least three days. If possible, get a prescription refill before the emergency occurs.

Veterinary Care: Know the location of the nearest emergency vet clinics and have their contact information available. If your pet has a chronic illness or specific health concerns, ensure you have documentation and emergency contacts for their care.

Calming Techniques: In stressful situations, pets can become anxious or frightened. To help manage their stress:

- Use calming collars or sprays (such as pheromone diffusers)
- Provide a quiet, familiar space for your pet to retreat to
- Keep your pet's routine as normal as possible during the crisis

Caring for Children During Emergencies

Children, especially younger ones, can easily become overwhelmed by the stress and uncertainty of an emergency. In addition to their safety and medical needs, children require emotional support and constant reassurance to help them cope with the situation.

1. Preparing Children for Emergencies

Teaching children how to react during an emergency is critical. While young children may not be able to execute all aspects of an emergency plan, they should be familiar with the basic concepts, such as:

- What to do if the fire alarm goes off
- Where the family meeting point is located
- How to contact an out-of-area relative
- For younger children, try to incorporate these lessons into play or storytelling to make them more relatable and less frightening.

Emergency Kit for Children: Your child's emergency kit should include:

- Enough food and water for at least 72 hours
- Medications, if applicable
- A comfort item, such as a favorite stuffed animal, blanket, or book, to help them feel secure
- A list of emergency contacts and any important documents (e.g., medical records, allergies)
- A flashlight with extra batteries

2. Emotional Support and Reassurance

Children may not fully understand the severity of a situation, but they can sense stress and fear. Emotional support is just as important as physical safety during an emergency. Keep the following in mind:

- **Routine and Familiarity:** Keep as much of their daily routine intact as possible. Familiar routines can help children feel more in control, which can reduce anxiety.
- **Answer Questions Simply:** Provide age-appropriate explanations about the emergency situation. Avoid overwhelming them with too much information but ensure they understand why they need to follow the emergency plan.

- **Stay Calm and Reassuring**: Children take emotional cues from adults, so try to stay calm. If you show panic or anxiety, it's more likely that they will become fearful and unsettled. Reassure them that you are taking steps to keep them safe.

3. Special Considerations for Infants and Toddlers

If you have an infant or toddler, you will need to address specific needs such as feeding and comfort:

- **Baby Formula and Diapers**: Stock up on extra baby formula, bottles, diapers, wipes, and other essentials. Make sure to have a way to sterilize bottles and pacifiers if necessary.
- **Sleeping Arrangements**: Infants and toddlers may require a familiar crib or sleeping setup. Consider a portable crib or sleep space that can be easily transported.

Caring for the Elderly During Emergencies

Elderly individuals often face unique challenges during emergencies due to health conditions, mobility issues, and cognitive impairments. Whether they live with you or you are caring for elderly relatives, it's crucial to prepare for their specific needs.

1. Special Health Considerations

Many elderly individuals rely on medical equipment or have specific health conditions that need attention during an emergency. The following considerations are crucial:

- **Medications**: Ensure a supply of prescription medications, along with a list of the medications they take, dosage information, and their prescribing doctor's contact information. Ensure the medications are in their original containers with labels visible.
- **Mobility Aids**: If the elderly person requires mobility aids (e.g., walkers, canes, wheelchairs), make sure these are easily accessible during an evacuation.
- **Oxygen Tanks and Other Medical Devices**: If your loved one requires supplemental oxygen or any other specialized medical equipment, ensure you have backup supplies, such as portable oxygen tanks or batteries for medical devices. Know how to safely transport these items.

2. Cognitive or Sensory Impairments

Some elderly individuals may have cognitive impairments (e.g., dementia) or sensory issues (e.g., hearing loss or vision impairments). These conditions require special considerations:

- **Familiar Reminders**: Label rooms and items within the house to help elderly individuals navigate if they become disoriented.
- **Communication Assistance**: If your loved one has difficulty hearing or seeing, make sure to provide tools to help them understand emergency alerts, such as a loud emergency alarm system, flashing lights, or visual cues.
- **Simplify Instructions**: In cases of cognitive impairment, provide clear, simple instructions and offer constant reassurance.

3. Comfort and Emotional Support

The elderly may experience heightened anxiety during an emergency, especially if they have pre-existing health conditions or mobility challenges. Offer reassurance by:

- **Maintaining Familiarity:** Surround them with familiar objects or comfort items, such as family photos or blankets, to reduce stress.
- **Comforting Presence:** Provide emotional support by remaining close and maintaining a calm demeanor. Physical touch, such as holding hands, can also be a powerful way to provide comfort during uncertain times.
- **Keeping Them Informed:** Give regular updates about the situation in a calm, reassuring manner. This can help them feel more in control and reduce anxiety.

In times of crisis, the well-being of pets, children, and the elderly must remain a top priority. By preparing ahead of time and ensuring that their unique needs are addressed, you can reduce their stress and increase their chances of staying safe and healthy during emergencies. When creating your emergency preparedness plan, always remember that your loved ones come in many forms—pets, children, and elderly relatives—and each one deserves careful attention and support. By equipping yourself with the knowledge, resources, and tools to care for them, you ensure that every member of your household is protected and cared for, no matter the circumstances.

5. HOME DEFENSE AND SECURITY

Fortifying Doors, Windows, and Entry Points

In a world where natural disasters, civil unrest, or other emergencies can happen at any time, securing your home becomes a top priority for ensuring your safety and that of your loved ones. Fortifying doors, windows, and entry points is an essential part of preparing your home for these situations. Whether you're dealing with the possibility of break-ins, storm damage, or civil disorder, making sure your home's entry points are secure will give you peace of mind and improve your chances of staying safe. In this subchapter, we will explore different ways to fortify these vulnerable areas of your home to ensure they can withstand a variety of threats.

1. Fortifying Doors

The Importance of Strong Doors

The front door is the primary point of entry for anyone attempting to gain access to your home. Whether you are concerned about burglary or simply want to protect your family during an emergency, having a strong, secure door is your first line of defense.

1.1 Choose the Right Door Material

Not all doors are created equal, and it's essential to choose a material that offers both security and durability. Some of the most secure materials include:

Solid Wood Doors: A solid wood door can provide a sturdy defense against forced entry. Look for doors made from hardwoods such as oak, mahogany, or hickory, which are more difficult to break through compared to softer woods like pine.

Steel Doors: Steel doors are one of the most robust options available. They offer superior resistance to break-ins and are less prone to warping or swelling from moisture. Steel doors are often reinforced with a foam core for added strength.

Fiberglass Doors: Fiberglass doors are also a good choice for home fortification, especially if you're looking for a door that combines strength with energy efficiency. They are resistant to dents and can be reinforced with steel or wood cores to enhance security.

1.2 Reinforcing the Door Frame

The strength of the door itself is important, but the door frame is often the weak link. Reinforcing the door frame will help ensure that it can withstand heavy pressure and prevent the door from being kicked in.

Use Steel Reinforcements: Install a steel reinforcement plate along the sides and top of the door frame. This helps to strengthen the frame and prevent it from splintering under impact.

Install a Door Jamb Reinforcer: A door jamb reinforcer, which is made of steel, is attached around the perimeter of the doorframe to strengthen it. This makes it more difficult for intruders to break in using force.

1.3 High-Security Locks and Hardware

Even the strongest door will be ineffective without high-quality locks. Install a deadbolt lock with a solid steel bolt that extends deep into the door frame. Use the following types of locks for maximum security:

- **Deadbolt Locks:** A single-cylinder deadbolt is a standard choice for residential security. It's operated with a key on the outside and a thumb-turn on the inside. Ensure the deadbolt extends at least one inch into the doorframe.
- **Multipoint Locks:** A multipoint locking system operates on the principle that several locking points engage simultaneously when the door is closed, making it much harder to force open.
- **Slide Bolts and Security Bars:** In addition to your regular lock, consider adding slide bolts or a security bar at the top or bottom of the door. This will prevent the door from being forced open, even if the lock is compromised.

1.4 Peepholes and Surveillance Systems

A peephole or wide-angle lens installed in your door allows you to see who's outside before opening it. This simple addition can prevent unwanted visitors from gaining access.

Adding a video surveillance system or a smart doorbell with a camera and two-way audio gives you the ability to see and communicate with anyone at your door, whether you're inside your home or elsewhere.

2. Fortifying Windows

Windows are one of the most vulnerable points of entry for burglars or intruders during an emergency. Fortifying your windows will help prevent unauthorized entry and reduce the risk of damage from environmental forces like strong winds or flying debris during a storm.

2.1 Reinforcing Window Glass

Security Window Film: Applying a security film to your windows is one of the most cost-effective ways to reinforce them. This film makes the glass shatter-resistant, helping to prevent windows from breaking easily. It also reduces the likelihood of flying glass during a storm or break-in attempt.

Laminated Glass: Laminated glass, often referred to as "safety glass," is an ideal option for homeowners who want additional protection. This type of glass consists of two panes of glass with a plastic layer sandwiched in between, making it harder to break.

Shatterproof Glass: Shatterproof or tempered glass is designed to resist impacts and will not break into sharp pieces if shattered. While it's more expensive than regular glass, it provides excellent protection against break-ins and environmental factors.

2.2 Installing Window Bars or Grills

While window bars might not be aesthetically pleasing, they provide an excellent deterrent against break-ins. Window bars can be custom-designed to fit the style of your home while offering robust protection.

Fixed Bars vs. Removable Bars: Fixed bars are bolted into the window frame and cannot be removed from the outside. However, removable bars are a more flexible option, offering protection while still allowing for exit during an emergency.

Decorative Grills: For homeowners who want to maintain a more attractive look, decorative window grills can provide both style and security. These grills can be installed on the outside of the windows and made from materials like steel or wrought iron.

2.3 Reinforced Window Frames and Locks

Adding window locks and reinforcing window frames can help protect your home from forced entry. Window locks prevent the window from being opened unless the key or code is used, making it difficult for intruders to simply slide the window open.

Consider reinforcing your window frames with steel or aluminum plates to make them more resistant to impact. Also, adding security bolts can make it nearly impossible to slide the window open from the outside.

3. Securing Other Entry Points

Your home may have other potential entry points such as sliding glass doors, garages, or attic hatches. Fortifying these areas is just as important as reinforcing your main doors and windows.

3.1 Sliding Glass Doors

Sliding glass doors are convenient but vulnerable to break-ins, as they are easy to bypass with a simple crowbar. Consider these security measures:

Sliding Door Lock Bar: Install a steel or aluminum security bar in the track of the sliding door. This prevents the door from being lifted off its tracks or forced open.

Glass Break Sensors: Install sensors that detect the sound or vibration of breaking glass. These systems are part of a home security system and can alert you and the authorities immediately if the door is breached.

Security Film or Shatterproof Glass: Apply the same security film or use laminated glass for sliding doors as you would for regular windows to reduce the risk of glass shattering during a break-in attempt.

3.2 Attic and Basement Hatches

In addition to the main entrance and windows, your home may have attic and basement hatches that can be used as entry points. Fortifying these areas is important:

Reinforced Hatch Covers: For attic and basement hatches, use steel or heavy-duty wood covers that can be locked into place. Consider installing security bolts or a heavy-duty deadbolt for additional protection.

Surveillance Cameras: Install security cameras to monitor these less-visible areas. Motion-activated cameras can alert you to any suspicious activity near these entry points.

Fortifying doors, windows, and entry points is a crucial aspect of securing your home during an emergency. Whether you are preparing for natural disasters, civil unrest, or simply a break-in, these protective measures can give you the peace of mind that your home and family are safe. By investing in high-quality materials, employing security systems, and reinforcing key entry points, you significantly reduce your risk of harm or loss. Your home is your sanctuary, and with the right precautions, it can stand strong against any threat that arises.

DIY Home Surveillance Systems

In an era where security threats are more prevalent than ever, homeowners are increasingly turning to DIY home surveillance systems to protect their property, loved ones, and belongings. Whether you're concerned about break-ins, intruders, or simply want to monitor activity around your home, a DIY surveillance system offers a cost-effective and customizable solution. These systems can be installed and managed without professional help, giving you complete control over your security. In this subchapter, we will explore the components of a DIY home surveillance system, the benefits of building your own system, and how to set up a system that meets your needs.

1. Understanding the Basics of DIY Home Surveillance Systems

Before diving into the specifics of setting up a surveillance system, it's important to understand the core components that make up a home security system. A DIY home surveillance system typically consists of the following elements:

- **Cameras:** Cameras are the core of any surveillance system. They capture real-time footage of your property, both indoors and outdoors. They come in a wide variety of types, each designed for different applications.
- **Storage:** Storage is crucial for saving video footage. Depending on your system, footage can be stored locally on a DVR (digital video recorder) or NVR (network video recorder), or in cloud storage for easy remote access.
- **Monitors/Viewing Devices:** Monitors or viewing devices allow you to access and view live or recorded footage from your cameras. These devices can include anything from a simple TV screen to a mobile phone or tablet with a dedicated app.

- **Sensors and Alarms:** Many surveillance systems also integrate motion detectors, door/window sensors, and alarms that trigger when the system detects unusual activity.
- **Connectivity:** A DIY surveillance system can be connected either through a wired or wireless connection. Wireless systems offer the flexibility of placement and are easier to install, while wired systems typically provide more stable connections.
- **Smart Integration:** Many DIY systems are now integrated with smart home technology, allowing you to control and monitor the system remotely via smartphones or voice commands using Alexa, Google Assistant, or other smart home platforms.

2. Benefits of a DIY Home Surveillance System

2.1 Cost-Effectiveness

One of the most attractive features of DIY home surveillance systems is their cost-effectiveness. When compared to hiring a professional security service, a DIY system typically saves you money on installation fees, monthly service charges, and equipment costs. DIY systems often come in packages with everything you need to set up your security system, or you can choose to customize your setup based on your specific needs and budget.

2.2 Customizability

A DIY home surveillance system allows you to customize the setup based on your unique security needs. Whether you live in a small apartment or a large home, you can choose the number of cameras, types of cameras, storage solutions, and monitoring methods that best suit your space. This flexibility allows you to scale your system over time as your needs evolve.

2.3 Full Control and Monitoring

With a DIY system, you have complete control over your home security. You can monitor the system remotely from your smartphone or computer, adjust settings, and even integrate the system with other smart home devices. This gives you peace of mind, as you're not relying on a third party to manage your security.

2.4 Easy Installation and Setup

Many DIY systems are designed for easy installation, even for people with little to no technical expertise. Step-by-step instructions, user-friendly apps, and plug-and-play components make it possible for homeowners to set up their surveillance system without the need for a technician. This convenience is one of the reasons why DIY systems are becoming more popular.

2.5 Expandability

As your security needs change, you can expand your DIY surveillance system with ease. Whether you want to add more cameras, sensors, or integrate new technology, DIY systems are built to accommodate future upgrades and modifications without requiring a complete system overhaul.

3. Choosing the Right Components for Your DIY Surveillance System

3.1 Cameras

Cameras are the most important part of any surveillance system. When choosing the right camera for your system, consider the following:

- **Resolution:** The resolution of a camera determines the clarity and detail of the video. For home security, cameras with at least 1080p resolution (HD) are recommended. Higher resolution cameras, such as 4K, offer even better detail but may be overkill unless you have large areas to monitor.
- **Indoor vs. Outdoor Cameras:** Outdoor cameras are built to withstand the elements, including rain, wind, and extreme temperatures. Make sure that the cameras you choose are weatherproof and designed for outdoor use if you plan to monitor the exterior of your home.
- **Wired vs. Wireless:** Wireless cameras are easier to install because they don't require complex wiring, but they depend on Wi-Fi signals for operation. Wired cameras offer more stability but are harder to install. Consider your space and the type of installation you're comfortable with when choosing between wired and wireless.
- **Field of View:** A camera with a wider field of view (FOV) can cover more area with fewer cameras. Look for cameras with a wide-angle lens, ideally around 120-180 degrees.
- **Night Vision:** If you plan to monitor your property at night, ensure that the cameras you choose have infrared (IR) night vision. This allows cameras to capture clear footage in low light conditions.
- **Motion Detection:** Motion detection sensors can trigger alerts and begin recording when they detect movement in the camera's field of view. This feature is crucial for maximizing the effectiveness of your system and conserving storage.

3.2 Storage

Storage is critical for keeping recorded footage, and you have several options to choose from:

Cloud Storage: Cloud-based storage allows you to store footage remotely on secure servers. Many DIY systems offer free or subscription-based cloud storage options, which come with varying amounts of storage space.

Local Storage (DVR/NVR): A more traditional method is to use a DVR or NVR system, which stores footage on hard drives. This option requires more physical space for installation but gives you greater control over your data.

Hybrid Systems: Some systems offer hybrid storage solutions that combine cloud and local storage, providing both remote access and physical storage backup.

3.3 Sensors and Motion Detectors

In addition to cameras, many DIY surveillance systems also integrate sensors and motion detectors that can trigger alarms and send alerts to your phone when activity is detected. Common sensors include:

Door and Window Sensors: These sensors detect when a door or window is opened and send an alert. They are often paired with other components like cameras or alarms.

PIR (Passive Infrared) Sensors: PIR motion sensors detect movement through heat signature changes. These sensors are ideal for monitoring entry points and large outdoor areas.

Glass Break Sensors: These sensors detect the sound frequency of glass breaking, providing additional protection for windows and glass doors.

3.4 Integration with Smart Home Devices

One of the biggest advantages of DIY surveillance systems is the ability to integrate them with other smart home devices. Many systems now allow for integration with:

Smart Thermostats: Some security systems can adjust your smart thermostat settings based on security activity, such as lowering the temperature when the system is armed.

Smart Lighting: Automatically turn lights on or off based on camera detection or schedule lighting to simulate activity when you're away from home.

Smart Locks: Integrating cameras with smart locks allows you to remotely unlock doors and control access to your home.

4. Installation and Setup Tips

4.1 Plan Your Camera Placement

The key to a successful DIY home surveillance system is strategic placement of cameras. Consider the following:

Entrances and Exits: Install cameras at all potential entry points, including front doors, back doors, and side windows.

High-Traffic Areas: Place cameras in areas that are commonly used by family members or intruders, such as hallways, driveways, and garage doors.

Outdoor Coverage: Ensure that your outdoor cameras cover areas like the driveway, yard, and any pathways leading to your home.

Avoid Blind Spots: Avoid placing cameras where obstructions, such as trees or furniture, could block the view. Ensure your cameras have a clear line of sight to important areas.

4.2 Test Your System

Once your system is installed, it's essential to test each component to ensure everything is working as it should. Test the cameras, motion detectors, and sensors, and check the video feed to ensure the footage is clear and the system is operating as expected.

4.3 Regular Maintenance

A DIY surveillance system requires ongoing maintenance to ensure it remains functional. Regularly check the camera lenses for dirt or debris, test the sensors, and ensure that the storage devices are not full or corrupted.

DIY home surveillance systems offer an effective, customizable, and affordable way to protect your home from potential threats. By selecting the right cameras, storage solutions, and integrating smart devices, you can create a security setup that fits your needs and budget. With the ability to monitor your property remotely, maintain full control over your security, and expand as needed, a DIY surveillance system is an essential component of any modern home defense plan.

Creating a Safe Room for Your Family

In an uncertain world, one of the most important preparations a family can make is creating a safe space within the home where they can take refuge in the event of an emergency. Whether it's a natural disaster, home invasion, civil unrest, or other emergency situations, a well-designed safe room can be the difference between life and death. In this subchapter, we will explore the essential components of creating a safe room, how to prepare it, and how to ensure that it functions effectively when needed most.

1. What is a Safe Room?

A safe room is a fortified, secure area within a home that is designed to provide shelter during dangerous situations. It is a designated space where family members can retreat to protect themselves from threats, including severe weather, intruders, and other emergencies. Safe rooms are typically equipped with necessities to ensure the occupants can remain safe and self-sufficient for a period of time until the threat has passed.

The key features of a safe room are its ability to:

- Withstand external threats such as intruders, gunfire, or natural disasters (like tornadoes or hurricanes).
- Provide a secure, comfortable environment for the duration of an emergency.
- Contain essential supplies and communication tools to keep the family protected and informed.

2. Assessing Your Needs and Threats

Before you begin building or setting up your safe room, it's essential to assess the risks and determine what kind of threats your family might face. The nature of the threat will influence the design, location, and equipment needed for the room.

2.1 Identifying Potential Threats

Some of the most common threats that might require a safe room include:

- **Home Invasion:** This could be due to criminal activity, civil unrest, or even domestic threats.
- **Natural Disasters:** Tornadoes, earthquakes, and hurricanes can devastate a home, and a safe room can provide shelter against such catastrophes.
- **Active Shooter or Terrorist Threat:** In some locations, especially urban areas or places at risk for terrorism, having a fortified room can save lives during a shooting or terror attack.
- **Chemical or Biological Attacks:** In rare but extreme cases, chemical or biological warfare can pose a threat, requiring a room sealed against contaminants.

Identifying the most likely risks in your area, as well as personal circumstances (such as children, elderly family members, or pets), will guide your decisions on where to build the safe room and what features to include.

3. Choosing the Location

The location of your safe room is one of the most critical factors. Ideally, it should be easily accessible, secure, and centrally located. Consider the following:

3.1 Central Location

The safest place for a safe room is typically in the center of the home. This minimizes the chances of external threats such as intruders or the structural damage that can occur with natural disasters.

Locations such as basements or interior rooms without windows, like closets, bathrooms, or pantries, can be ideal. A basement, in particular, offers better protection against tornadoes and severe storms.

3.2 Accessibility

The room should be easily accessible to all family members, including children and elderly individuals. This means that it should be free from obstructions and located within a reasonable distance from common living areas.

Avoid placing the safe room on the upper floors of a building, as these areas are more vulnerable to damage in events like earthquakes or fires.

3.3 Concealment

Ideally, a safe room should be discreet and not easily visible to potential intruders. You can build a safe room behind a false wall, within a basement, or even in a closet that's camouflaged as a storage area.

The room should be designed in such a way that it isn't obvious to someone searching your home.

4. Reinforcing the Room for Security

The core purpose of a safe room is to keep you protected, and its construction should reflect that. Reinforcing the room and its entry points is a priority. Here's how to do it:

4.1 Reinforced Walls

- **Materials:** The walls should be made of materials that can withstand significant force, including reinforced concrete, steel, or ballistic-resistant panels. In some cases, you might consider using a steel door with a concrete or cinder block wall behind it.
- **Bulletproofing:** If your area faces a risk of armed threats, bulletproof materials such as Kevlar, steel, or other ballistic-resistant options can be used for the door and walls to offer protection from gunfire.

4.2 Door and Locking Mechanism

The door is one of the most vulnerable parts of a safe room. It should be equipped with a heavy-duty steel door with a high-quality locking mechanism. Consider reinforced steel doors that can withstand kicks or forced entry.

A biometric lock or keypad system can be an added security feature, allowing only authorized users to access the safe room.

4.3 Windows and Ventilation

Ideally, a safe room should have no windows to reduce the chances of being exposed to an attack. If windows are present, ensure they are bulletproof or have reinforced coverings to protect from flying debris or gunfire.

Proper ventilation is also crucial, especially if the room needs to be sealed in the case of a biological or chemical attack. Ventilation systems that use filters can keep the air fresh and safe. It's vital that the system doesn't allow outside contaminants to enter the room.

5. Essential Supplies for Your Safe Room

Once you have chosen and reinforced your location, it's time to think about what supplies you'll need to sustain yourself and your family. A well-stocked safe room should include the following items:

5.1 Emergency Food and Water

Water: Store enough water for every member of the household to survive for at least 72 hours. A general guideline is one gallon per person per day for drinking, cooking, and hygiene.

Food: Non-perishable, high-calorie foods such as canned goods, freeze-dried meals, and energy bars should be included. Store enough for 72 hours or longer.

5.2 Medical Supplies

First Aid Kits: Include basic first aid supplies, such as bandages, antiseptics, painkillers, and specialized medication (e.g., insulin for diabetic family members).

Prescription Medications: Ensure that essential medications are stored in the safe room.

CPR Masks and Emergency Kits: For situations that require advanced first aid.

5.3 Communication Devices

Two-Way Radios: A reliable communication system is essential. A battery-powered or hand-crank radio can help you stay informed if you lose access to other communication methods.

Cell Phones with Extra Chargers: Keep cell phones in the safe room, and make sure they're charged or have backup power sources (e.g., portable battery packs).

Satellite Phones: In remote areas, or if you expect your regular cell network to be disrupted, consider keeping a satellite phone for emergency communication.

5.4 Lighting and Power

- **Flashlights:** Keep multiple flashlights and spare batteries, as power outages are likely during an emergency.
- **Battery-Powered or Solar-Powered Lights:** Having backup lighting can provide comfort and reduce the stress of an emergency.

5.5 Defense Tools

Firearms: In some situations, you may need to defend yourself. Having firearms, ammunition, and knowledge of how to use them responsibly is a personal choice for many families.

Pepper Spray or Mace: Non-lethal tools for protection against intruders.

5.6 Comfort and Personal Items

- **Blankets and Pillows:** Comfort is essential during a prolonged stay in the safe room.
- **Personal Documents:** Store copies of vital documents such as IDs, passports, medical records, and insurance papers.
- **Toys and Games for Children:** If children are in your household, bring items to keep them entertained and calm.

6. Training and Drills

Creating a safe room is only part of the process. It's essential that every family member is trained on how to use the room and the tools within it.

6.1 Regular Drills

Conduct regular family drills to ensure everyone knows where the safe room is located, how to access it, and how to use the equipment inside. Practice sealing the room and remaining inside for an extended period to simulate real-life conditions.

6.2 Emergency Plans

Develop emergency plans, including escape routes, communication strategies, and a family reunion point if the safe room becomes inaccessible.

A safe room is an essential component of any modern home defense strategy. It provides peace of mind, knowing that you and your family have a secure space to retreat to in times of crisis. By carefully selecting the location, reinforcing the room, and stocking it with the necessary supplies, you can ensure that your safe room serves as a stronghold in times of emergency. While building and maintaining a safe room may seem like a large investment of time and resources, the safety and security it provides are invaluable.

Defensive Landscaping for Added Protection

In today's world, where the threat of natural disasters, civil unrest, and even home invasions is a real concern, creating a secure environment around your home is just as important as reinforcing the structure itself. One of the most effective yet often overlooked aspects of home security is defensive landscaping. By using plants, trees, and other landscape features strategically, you can not only enhance the aesthetic appeal of your property but also improve its safety and defensibility.

Defensive landscaping refers to the use of natural elements to deter, delay, or prevent unwanted access to your home or property. Whether you are looking to protect your home from burglars, limit the impact of a natural disaster, or create a safer environment during an emergency, incorporating defensive landscaping techniques can play a significant role in achieving these goals.

In this section, we will explore the concept of defensive landscaping, the benefits of various plants and design techniques, and how to create a protective, defensible perimeter around your home.

1. Understanding the Purpose of Defensive Landscaping

The primary objective of defensive landscaping is to create a barrier between your home and potential threats, making it more difficult for intruders or unwanted individuals to approach or enter your property. By employing the right combination of plants, hedges, trees, and hardscape elements, you can:

- **Deter Criminals:** Dense hedges, thorny plants, and strategically placed trees can act as physical barriers that make it more difficult for intruders to access windows, doors, or areas of your home.
- **Slow Down Intruders:** A well-designed landscape can help slow down an intruder's movement, buying you precious time to alert authorities or escape if necessary. This can include using thick hedges, rocks, or other obstructions to create an obstacle course for potential threats.
- **Provide Privacy and Concealment:** Dense foliage can shield your home from view, giving you privacy while also reducing the chance that intruders will know when you are home or see vulnerable entry points.
- **Offer Natural Surveillance:** By using landscaping strategically, you can create sightlines that allow you to spot potential threats or suspicious activity before it escalates. This could involve positioning bushes or trees in areas that maximize visibility from windows or outdoor areas.

2. Choosing the Right Plants for Defensive Landscaping

The plants you choose for defensive landscaping play a significant role in both aesthetics and security. Some plants have natural deterrent properties, while others are primarily used for their ability to slow down or obstruct an intruder's movement. Below are some of the best plants and trees to incorporate into your defensive landscaping plan:

2.1 Thorny Shrubs and Bushes

Thorny plants are some of the most effective in defensive landscaping because they create a natural barrier that can make it difficult for intruders to approach without injury. Common examples of thorny shrubs and bushes include:

- **Holly (Ilex aquifolium):** Known for its sharp, spiky leaves, holly is an excellent choice for creating a natural barrier around windows and fences. It also provides year-round coverage with its evergreen leaves.
- **Barberry (Berberis vulgaris):** Barberry shrubs are armed with sharp thorns that make them a formidable deterrent against anyone trying to pass through them. They also offer vibrant colors, making them both functional and visually appealing.
- **Crown of Thorns (Euphorbia milii):** This thorny plant has long, sharp thorns and is commonly used as a deterrent for fences and perimeters. It's a low-maintenance option that adds beauty and security.
- **Roses (Rosa spp.):** Certain types of roses, particularly climbing varieties with thick thorns, can make it difficult for intruders to climb over fences or walls, as the thorns will cause discomfort.

2.2 Dense, Thick Hedges

Dense, fast-growing hedges provide both security and privacy, effectively blocking access to windows and doors. These can be used to form a living fence or wall around your property. Some of the best hedge plants for defensive landscaping include:

- **Privet (Ligustrum spp.):** Privet hedges are quick-growing, dense, and can be easily shaped to form solid walls. They're ideal for creating privacy while also providing a physical barrier to prevent entry.
- **Boxwood (Buxus spp.):** Boxwood is another excellent option for creating dense, low-maintenance hedges around the perimeter of your home. It can be pruned to maintain a specific height and shape, making it both practical and attractive.
- **Yew (Taxus spp.):** Yew hedges are slow-growing but dense and tough, making them a long-lasting addition to your landscaping. They are also evergreen, providing year-round protection.

2.3 Climbing Plants for Fences

Using climbing plants on fences and trellises can help create an additional layer of defense around your property. These plants are often difficult to scale because of their sharp leaves or thorny vines. Some climbing plants to consider include:

- **English Ivy (Hedera helix):** English Ivy grows quickly and can cover a fence or wall, making it harder for an intruder to climb over. The thick coverage also provides privacy and adds beauty to the landscape.
- **Clematis (Clematis spp.):** With some species having sharp, rough vines, clematis can climb fences and trellises, creating an attractive but difficult-to-navigate barrier.

- **Wisteria (Wisteria spp.):** While beautiful in appearance, wisteria is also tough to remove once it has established itself. Its woody vines make it challenging for an intruder to climb.

2.4 Trees for Defensive Landscaping

While trees can offer aesthetic beauty, they also provide multiple benefits in terms of defense. Certain trees can act as natural barriers and offer protection from the elements. Recommended trees for defensive landscaping include:

- **Hawthorn (Crataegus spp.):** Known for its spiky branches and dense foliage, hawthorn trees are great for creating a natural perimeter that deters intruders.
- **Alder (Alnus spp.):** Alder trees can be grown in clusters to create a thick, natural wall. They are also resistant to many pests and grow rapidly.
- **Maple (Acer spp.):** Maples are large, sturdy trees that can act as a strong physical deterrent. Their dense canopy also provides cover and adds privacy to your home.

3. Landscaping Features Beyond Plants

While plants are the core component of defensive landscaping, there are additional design elements you can incorporate to enhance the security of your property.

3.1 Gravel and Rock Pathways

Gravel or rock pathways around your home can serve as an excellent deterrent to intruders, as the sound of crunching gravel can alert you or others to someone's presence. Additionally, the uneven surface can make it difficult for individuals to approach your home quietly.

3.2 Perimeter Fencing

While not technically landscaping, a strong, high-quality fence can act as a solid line of defense. Combine this with climbing plants and thorny bushes to add an extra layer of security.

3.3 Water Features

Adding a small pond, stream, or fountain to your property can help slow intruders down by creating an obstacle that they must navigate. This not only adds beauty but can also reduce the ease with which someone can enter your property.

4. Maintenance of Defensive Landscaping

To ensure that your defensive landscaping remains effective over time, regular maintenance is essential. This includes:

- **Pruning and Shaping:** Regularly trim hedges, shrubs, and trees to keep them healthy and dense. A well-maintained landscape is more secure and will serve as a stronger deterrent to potential intruders.

- **Replacing Damaged Plants:** If any plants are damaged, replace them quickly to maintain a continuous barrier. The effectiveness of your defensive landscaping depends on having a solid, uninterrupted line of defense.
- **Monitor Growth:** Some plants grow very quickly, and you will need to keep them in check so they don't obstruct entry points or pathways.

Defensive landscaping is an invaluable tool in creating a more secure home. By strategically using thorny plants, dense hedges, climbing vines, and trees, you can significantly enhance the protection of your property against a variety of threats. Whether you're trying to prevent a break-in, provide privacy, or create a safer environment during a natural disaster, defensive landscaping can be an effective and low-cost addition to your overall home security strategy. With the right combination of plants, design features, and consistent maintenance, you can create a landscape that not only enhances your home's appearance but also serves as a formidable defense against threats.

6. STOCKPILING AND RESOURCE MANAGEMENT

Food Storage Essentials: Short and Long-Term

When it comes to preparing for emergencies or simply ensuring that your household is self-sufficient in times of need, food storage is one of the most critical aspects of your survival plan. Whether you are dealing with natural disasters, supply chain disruptions, or other unforeseen circumstances, having a reliable stockpile of food is essential. In this subchapter, we will discuss the essentials of food storage, focusing on both short-term and long-term needs, and provide practical advice on how to store food effectively to ensure its longevity and safety.

1. Understanding the Importance of Food Storage

Food storage serves multiple purposes in a survival context:

- **Preparedness for Emergencies:** Natural disasters, power outages, civil unrest, and pandemics can disrupt the availability of food. By maintaining a well-organized food supply, you ensure that your family has access to nourishment when access to fresh groceries becomes difficult.
- **Self-Sufficiency:** For those who desire independence from grocery stores, a well-stocked food pantry allows you to be less reliant on external sources, ensuring that you can provide for yourself and your loved ones in case of a crisis.
- **Cost Savings:** Buying in bulk and storing food allows you to take advantage of sales, buy non-perishable items at a discount, and avoid the high costs of convenience foods, making it an economical choice over time.

2. Short-Term Food Storage

Short-term food storage refers to food supplies that are intended to be used within a short period (a few weeks to a few months). These items are typically fresh or semi-perishable, and their storage requires careful monitoring of expiration dates and proper conditions to prevent spoilage. While many short-term foods can be purchased and stored with minimal preparation, they must be used before their shelf life expires.

2.1 Key Short-Term Food Storage Items

- **Canned Goods:** Canned vegetables, fruits, meats, beans, and soups are essential components of any short-term food supply. These items have a long shelf life (usually 1–5 years), but they must be rotated regularly to ensure freshness. Store them in a cool, dry place, and check for any signs of bulging, rust, or leakage, as these may indicate contamination or spoilage.
- **Frozen Foods:** If you have access to a freezer, frozen fruits, vegetables, meats, and ready-to-eat meals are excellent short-term storage options. However, frozen food is only viable as long as your freezer remains powered. In case of power outages, it's important to have contingency plans, such as coolers and ice, to maintain the integrity of your frozen supplies.
- **Dried Foods:** Items such as pasta, rice, and dried beans are perfect for short-term storage. These staples can be easily stored in sealed containers and have a shelf life of several months to a year.

Proper sealing of dried foods in airtight bags or containers helps prevent moisture and pests from ruining them.
- **Fresh Produce:** While fresh produce doesn't last long, it can still be part of a short-term storage plan. Keep your fridge stocked with fruits, vegetables, and dairy products, and plan your meals to ensure these items are used before they spoil. Additionally, consider freezing some fresh produce to extend their shelf life.

2.2 Best Practices for Short-Term Food Storage

- **Cool, Dry, and Dark:** Store all short-term food in a cool, dry place away from sunlight. Excess heat and humidity can cause food to spoil more quickly.
- **Rotation System:** Implement a "first in, first out" (FIFO) system where you use older items before newer ones. This ensures you are consuming food before it expires and reduces waste.
- **Sealed Containers:** For dried and pantry foods, using airtight containers or Mylar bags with oxygen absorbers can help extend the shelf life. Sealing foods properly also keeps pests like rodents and insects away.
- **Inventory Management:** Keep an inventory of your short-term supplies, noting expiration dates and quantities. This will help you manage your stock more efficiently.

3. Long-Term Food Storage

Long-term food storage is designed to ensure that you have a reliable supply of food that can last for several months or even years. This is particularly important for emergency situations where access to fresh food may be restricted, such as in the aftermath of a disaster or during a long-term power outage. Long-term storage foods typically have a shelf life of several years or longer and require specific conditions to maintain their quality.

3.1 Key Long-Term Food Storage Items

- **Freeze-Dried Foods:** Freeze-dried foods, such as fruits, vegetables, meals, and meats, are some of the best options for long-term food storage. Freeze-drying preserves the nutritional content and flavor of the food while reducing its weight and volume. These foods can last 25-30 years when stored in airtight, oxygen-free packaging. They only require rehydration with water, making them easy to prepare.
- **Dehydrated Foods:** Similar to freeze-dried foods, dehydrated foods like soup mixes, rice, and dehydrated fruits also offer excellent long-term storage. While they have a shorter shelf life than freeze-dried foods, dehydrated foods still last 5–10 years when stored correctly. They are also lightweight and easy to store.
- **Grains and Legumes:** Bulk grains (e.g., wheat, oats, corn) and legumes (e.g., lentils, chickpeas, dried beans) are essential for long-term storage because they are rich in calories, protein, and fiber. They typically last 10–30 years if stored in airtight containers or Mylar bags with oxygen absorbers.
- **Powdered Milk and Eggs:** Powdered milk and egg products can be stored for many years and are great sources of nutrition, especially for growing children and those with specific dietary needs. These products are available in large, sealed containers and are often used in emergency food kits.

- **Canned Goods (Extended Shelf Life):** While traditional canned goods are also valuable for long-term storage, you can find specialty long-term canned foods with extended shelf lives (10–25 years). These foods, such as meat, fruits, and vegetables, are often specially processed to last much longer than typical canned goods.
- **Honey, Sugar, and Salt:** These pantry staples have an indefinite shelf life when stored properly. They are ideal for long-term storage due to their preservative qualities and nutritional value in times of need.

3.2 Best Practices for Long-Term Food Storage

- **Temperature Control:** Store long-term foods in a cool, dry, and dark location. The ideal temperature for long-term food storage is around 50°F to 70°F (10°C to 21°C). Heat accelerates the degradation of food, reducing its shelf life.
- **Sealing and Packaging:** Proper sealing is critical for extending the shelf life of long-term foods. Use airtight containers, vacuum-sealed bags, Mylar bags, and oxygen absorbers to protect foods from oxygen, moisture, and pests.
- **Proper Inventory Management:** Long-term food storage requires careful monitoring. Maintain an inventory system to track the types, quantities, and expiration dates of your stock. Rotate foods periodically and replace items as needed to ensure your stock remains fresh.
- **Diversified Diet:** To ensure balanced nutrition during emergencies, stock a variety of food types. This includes fruits, vegetables, proteins, grains, dairy substitutes, and essential fats and oils. A well-rounded diet is crucial for maintaining health in the long term.

4. Considerations for Specialized Dietary Needs

When storing food for your family, it's essential to account for any dietary restrictions or preferences. If someone in your household has food allergies, intolerances, or special dietary needs, you'll need to select appropriate items that fit those requirements. For example:

- **Gluten-Free:** There are many gluten-free options available for both short-term and long-term storage, including gluten-free grains (quinoa, rice, oats) and prepared meals.
- **Vegan and Vegetarian:** Protein-rich vegetarian options such as dehydrated tofu, beans, legumes, and plant-based freeze-dried meals can help ensure a complete diet.
- **Dairy-Free:** Store non-dairy alternatives like powdered almond or coconut milk for individuals who are lactose intolerant or avoid dairy products.

Building a solid food storage plan, both short-term and long-term, is crucial for maintaining self-sufficiency and readiness in times of need. By understanding the differences between short-term and long-term food storage and following best practices for each, you can ensure that you have the food you need to weather any emergency. Whether it's keeping a few extra cans of soup on hand or investing in freeze-dried meals for years to come, having a well-organized and balanced food storage plan is a key part of your survival strategy.

Water Storage, Filtration, and Purification

Water is the most vital resource for survival, yet it is often one of the most overlooked aspects of emergency preparedness. In an emergency situation, access to clean drinking water can quickly become scarce, and knowing how to store, filter, and purify water could mean the difference between life and death. In this subchapter, we will explore the critical components of water storage, filtration, and purification, providing practical advice on how to ensure your family has access to safe, clean water when it's most needed.

1. The Importance of Water in Survival Situations

Water is essential for human survival, and the human body can only go without water for about 3 to 5 days depending on environmental conditions, physical activity, and individual health. However, dehydration can set in much sooner, and even mild dehydration can lead to serious health issues such as dizziness, confusion, and organ failure. In survival situations, where resources may be limited, having access to water is paramount.

Water is necessary for:

- **Hydration:** Drinking water is crucial for maintaining bodily functions such as temperature regulation, digestion, and circulation.
- **Food Preparation:** You'll need water to cook and clean food.
- **Hygiene:** Water is essential for personal hygiene, wound care, and cleaning.
- **Sanitation:** In a prolonged emergency, access to water for cleaning and sanitation is vital to prevent illness and disease.

Given the importance of water, it's essential to plan ahead and ensure you have the means to store, filter, and purify it, particularly in areas where water supply may be compromised due to natural disasters, power outages, or supply chain disruptions.

2. Water Storage for Emergencies

Water storage is one of the first steps in preparing for emergencies. Without a sufficient amount of stored water, you may quickly run out of this precious resource, especially if the emergency persists for several days or longer. Proper water storage requires careful planning, ensuring that you store an adequate supply for both drinking and other essential uses.

2.1 How Much Water to Store

The general recommendation for water storage is to store at least 1 gallon of water per person per day for drinking, cooking, and hygiene. For a family of four, this means at least 4 gallons per day. You should plan to store enough water to last at least 72 hours, though a week or more is ideal, depending on your circumstances.

For long-term emergencies, water usage may increase, and it's recommended to store 1 to 2 gallons per person per day for a longer period. You can also consider storing additional water for sanitation and hygiene purposes.

2.2 Best Practices for Water Storage

- **Use Proper Containers:** Store water in clean, food-grade containers made from BPA-free plastic, glass, or stainless steel. Avoid using containers that previously held chemicals, as these may contaminate the water.
- **Consider Water Storage Capacity:** Water is heavy (1 gallon weighs about 8.34 pounds), so you need to factor in both the weight of water and the space available for storage. Consider stacking 5-gallon water containers or using larger water barrels (50 gallons or more) if you have the space.
- **Rotate Water Supplies:** Water doesn't expire, but it can pick up contaminants if it's stored improperly. Be sure to rotate your water every 6 months to 1 year to keep it fresh. Mark your containers with dates to track when to refresh your supply.
- **Store Water in Multiple Locations:** To minimize the risk of losing your water supply in case of damage (e.g., leaks, spills), it's wise to store water in multiple locations within your home, or even outside, in a cool, dark, and protected environment.

2.3 Ideal Water Storage Containers

- **Water Barrels:** Typically available in 55-gallon sizes, water barrels are designed for long-term water storage. They should be food-grade and properly sealed to prevent contaminants.
- **Water Cans and Pouches:** These are smaller, portable water storage options that can be stacked and stored easily. They are often treated to extend shelf life and prevent bacterial growth.
- **Water Bottles:** While convenient for short-term use, standard bottled water isn't designed for long-term storage. However, it can be handy for quick access during the initial phases of an emergency.

3. Water Filtration

In emergency situations, water filtration is essential, especially when your primary water source is compromised or when you need to access water from questionable sources, such as lakes, rivers, or rainwater. Filtration removes contaminants, sediments, and particles from the water, making it safe to drink and use.

3.1 Types of Water Filters

- **Activated Carbon Filters:** These are commonly used for removing chlorine, volatile organic compounds (VOCs), and unpleasant odors from water. They can improve taste and make water more palatable.
- **Ceramic Filters:** Ceramic filters are effective at removing bacteria and protozoa, and they can last for many years if maintained properly. These filters can be used with a wide range of water sources, including streams, ponds, and lakes.

- **Reverse Osmosis Filters:** These filters are more expensive but highly effective at removing dissolved salts, minerals, and other contaminants. They require a water pressure source and can produce a higher quality of purified water.
- **Ultraviolet (UV) Filters:** UV filters use light to kill bacteria, viruses, and other pathogens in the water. They are easy to use and effective in providing purified water when used correctly.

3.2 Choosing the Right Filter

When selecting a water filter for your emergency preparedness kit, consider the following:

- **Portability:** If you are bugging out or need a mobile solution, choose a filter that is lightweight and easy to carry.
- **Filtration Capacity:** Consider how much water you need to filter in a given day. Some filters are designed for personal use (e.g., 1–2 liters per day), while others are larger and suitable for family-sized needs.
- **Maintenance:** Some filters, like ceramic filters, require regular cleaning to maintain their effectiveness. Consider the ease of maintenance when selecting your filtration system.
- **Effectiveness:** Not all filters are equally effective at removing all types of contaminants. Be sure to select a filter that targets the pathogens and pollutants most likely to be present in your local water sources.

4. Water Purification

While filtration removes physical contaminants and some pathogens, it doesn't always eliminate harmful microorganisms such as viruses or certain bacteria. This is where water purification comes in.

4.1 Methods of Water Purification

- **Boiling:** Boiling is one of the simplest and most effective methods of purifying water. Boiling for at least 1 minute (or 3 minutes at higher altitudes) will kill most bacteria, viruses, and parasites. However, it requires fuel and time, which may not always be available in an emergency situation.
- **Chemical Purifiers:** Common chemicals like iodine or chlorine dioxide can be used to purify water by killing bacteria, viruses, and other pathogens. These chemicals are compact and easy to store but may leave an aftertaste.
- **Water Purification Tablets:** These are pre-measured tablets that can be dropped into water to purify it. They are easy to carry and use, but they are effective only against specific types of contaminants.
- **Solar Purification:** This method, known as solar still or SODIS (Solar Water Disinfection), uses the sun's heat to evaporate and purify water. While it's a slow process, it's ideal when you have no other options available.

4.2 Choosing the Best Purification Method

The best water purification method depends on your specific needs, resources, and situation. If you are in a survival situation and have limited fuel or equipment, boiling or solar purification might be the most accessible options. If you have access to purification tablets or chemicals, these can be quicker and more efficient.

5. Combining Filtration and Purification

In many cases, it's advisable to combine filtration and purification to ensure your water is as safe as possible:

- **Filter First:** Use a water filter to remove large particles, sediment, and common pathogens.
- **Purify:** Follow up with a purification method, such as boiling or using purification tablets, to eliminate viruses and ensure the water is completely safe.

Water storage, filtration, and purification are foundational elements of any survival plan. By understanding the various methods of storing, filtering, and purifying water, you can ensure that you and your family have a consistent and safe water supply, regardless of the circumstances. Prepare in advance by storing enough water for at least 72 hours and implementing reliable filtration and purification systems. When disaster strikes, access to clean water could be the key to surviving until conditions improve.

Rotating and Preserving Supplies

One of the most crucial aspects of emergency preparedness is ensuring that your supplies remain fresh, functional, and ready for use when you need them the most. Rotating and preserving supplies is key to maintaining their quality, shelf life, and effectiveness. If you're relying on long-term storage for food, water, medical supplies, or other essential goods, it's vital to have a system in place to monitor, replace, and preserve these items to avoid waste and ensure preparedness in the event of an emergency.

1. The Importance of Rotating Supplies

Rotation is the process of using up older supplies first and replacing them with newer ones to prevent items from expiring, spoiling, or becoming obsolete. This is particularly critical for food and medical supplies, but it also applies to tools, batteries, and even survival gear.

Rotating supplies ensures:

- **Freshness and Quality:** Perishable goods, such as food and medicine, can degrade over time. Rotation helps maintain the quality of your supplies, ensuring that you always have usable products when needed.
- **Cost-Effectiveness:** By using older supplies first, you avoid waste and prevent unnecessary costs associated with discarding expired or expired items. It allows you to manage your budget by systematically replacing items as they approach their expiration dates.
- **Preparedness:** Emergency situations are unpredictable. By rotating your supplies regularly, you ensure that you are always prepared and that your stock is up to date, avoiding shortages or unavailability of crucial resources when a disaster strikes.

2. Strategies for Rotating Supplies

To effectively rotate your supplies, it's necessary to create an organized system that allows you to track the age, condition, and expiration dates of your goods. Here are several strategies for ensuring a smooth and effective rotation:

2.1 FIFO (First In, First Out)

One of the most commonly used methods for rotating supplies is FIFO, or First In, First Out. This system ensures that the oldest items are used first, which minimizes waste and keeps items fresh. To implement FIFO:

- **Label and Date:** As you purchase new supplies, label each item with the date of purchase and, if applicable, its expiration date. This can be done using a permanent marker or adhesive labels.
- **Organize by Age:** Place older items in the front of your storage area and newer items toward the back. This makes it easier to access the older items first.
- **Create Inventory Logs:** Maintain a log of your supplies and their expiration dates. This can be as simple as a written spreadsheet or using apps designed for inventory management. Apps like Pantry Check or My Pantry allow you to track expiration dates and set reminders for replacement.

2.2 The 6-Month Rotation Rule

For perishable items like food, it's essential to maintain a consistent rotation schedule. The 6-month rule is a good guideline for most non-frozen, non-refrigerated foods. Every six months, go through your stock to:

Use items with approaching expiration dates: Always consume or donate food products nearing their expiration.

Replace supplies: After using up old items, replace them with fresh stock.

Evaluate quality: Check all foods for signs of damage, such as dents, leaks, or tears in packaging, and inspect for any discoloration or mold. Discard any items that have gone bad.

2.3 Quarterly Inspections

In addition to the 6-month rotation rule, conducting quarterly inspections can help ensure your stock is maintained properly. This is particularly helpful for non-food items like medical supplies, batteries, and tools that don't necessarily expire but could become ineffective over time.

- **Check for damage:** Inspect all items for damage. This includes checking for rust, broken parts, and leaks in containers.
- **Test functionality:** Ensure that items like flashlights, radios, and emergency tools still work properly. Replace batteries if necessary.
- **Review conditions:** Consider temperature and humidity levels, as they can affect certain supplies like medications, freeze-dried food, and batteries.

3. Preserving Supplies for Long-Term Storage

In addition to rotating supplies, preserving them is another vital aspect of ensuring that your emergency stock remains viable for extended periods. Different types of supplies require different preservation methods. Here are key strategies for preserving various essential goods:

3.1 Food Preservation

When preparing long-term food supplies, it's important to store items in a way that minimizes spoilage and extends their shelf life. Some of the most effective methods include:

- **Freeze-Drying:** Freeze-dried foods can last for up to 25 years or more when stored in airtight containers. Freeze-drying removes the moisture from food, which helps prevent bacterial growth and spoilage.
- **Canning:** Canning involves sealing food in jars and then heating them to kill bacteria, yeasts, and molds. Properly canned goods can last for several years. Be sure to follow safe canning practices to avoid contamination.
- **Vacuum Sealing:** Vacuum sealing removes air from food packages, which helps prevent oxidation and keeps food fresh. It's particularly effective for meats, grains, and other dry goods. Vacuum-sealed items can last for months or even years depending on the type of food.
- **Mylar Bags with Oxygen Absorbers:** For long-term storage, foods like rice, beans, and grains can be sealed in Mylar bags with oxygen absorbers. This method prevents oxygen from damaging food and helps it last for years.

3.2 Water Storage Preservation

Water, unlike food, doesn't spoil in the traditional sense, but it can become contaminated if not stored properly. To preserve water:

- **Store in food-grade containers:** Only store water in containers that are specifically designed for water storage. These should be made of BPA-free plastic, glass, or stainless steel and be sealed tightly.
- **Use water preservatives:** If you're storing water for extended periods, consider using water preservatives such as potassium iodide or water treatment tablets to prevent bacterial growth.
- **Keep in a cool, dark place:** Water should be stored in a cool, dry, and dark area to prevent algae growth or contamination from sunlight.

3.3 Medicine and Medical Supplies Preservation

Preserving medical supplies is crucial for long-term preparedness, particularly since many over-the-counter medications have expiration dates. Here's how to extend their shelf life:

- **Store medications in a cool, dry place:** Heat, light, and moisture can degrade medications over time. Keep them in a temperature-controlled environment, ideally between 59°F and 86°F (15°C to 30°C).
- **Keep in airtight containers:** Use sealed, airtight containers to store medications and prevent moisture from getting inside.

- **Use desiccants:** Desiccant packets, which absorb moisture, can be placed inside medication containers to help keep them dry and fresh.

4. Preventing Common Rotating and Preservation Mistakes

While rotating and preserving supplies is important, there are some common mistakes that can compromise the effectiveness of your preparations. Be mindful of the following:

- **Not checking expiration dates:** It's easy to forget to check expiration dates or assume that an item will last forever. Be diligent about inspecting your supplies regularly.
- **Storing supplies improperly:** Excessive heat, humidity, or direct sunlight can quickly degrade the quality of your supplies. Ensure your storage space is cool, dark, and dry.
- **Overstocking and under-rotating:** Buying in bulk can be beneficial, but only if you rotate your stock effectively. Overstocking without regular rotation can lead to spoilage and waste.

Rotating and preserving supplies is an essential practice for anyone preparing for emergencies. By establishing a system for inventory management, using appropriate storage methods, and rotating supplies regularly, you can ensure that your family will be well-prepared, with access to fresh, usable resources when an emergency arises. Whether you're stockpiling food, water, medical supplies, or survival gear, it's crucial to maintain the integrity of these supplies through careful rotation and preservation. With these practices in place, you'll be ready for whatever challenges the future may bring.

Medical Kits and Specialized Tools

A well-prepared medical kit is a cornerstone of any survival plan, particularly for those opting to "bug in" during an emergency or disaster. In critical situations, access to professional healthcare can be limited or non-existent, making it essential to have the right tools and knowledge to address medical needs at home. A comprehensive medical kit not only helps in dealing with everyday injuries and illnesses but also ensures you're ready for more serious situations, whether it's natural disasters, pandemics, or other emergencies.

This chapter will explore the components of a well-stocked medical kit, including general and specialized tools that every prepper should consider. It will also highlight best practices for maintaining and organizing your medical supplies to ensure they're ready when you need them most.

1. The Essentials of a Basic Medical Kit

At the foundation of any survival medical kit is the ability to treat a range of common injuries and medical conditions. A well-rounded kit should be versatile and include supplies for wound care, pain management, infection prevention, and stabilization of vital functions in the event of a serious injury or illness. Here are some essential items that should be in every medical kit:

1.1 Wound Care Supplies

Bandages: A variety of sizes and types (adhesive, gauze, non-stick) for covering wounds. Band-aids, sterile gauze pads, and adhesive tape are fundamental.

- **Antiseptics and Disinfectants:** Alcohol swabs, iodine solutions, hydrogen peroxide, and antiseptic wipes to clean wounds and prevent infections.
- **Sterile Dressings:** For covering larger or more serious wounds to prevent infection.
- **Burn Care:** Gel or ointment for burns (like aloe vera or burn relief cream), along with sterile dressings to cover burns until they can be properly treated.
- **Cotton Balls and Swabs:** Used for cleaning and applying topical treatments.

1.2 Pain and Fever Management

- **Pain Relievers:** Over-the-counter medications like acetaminophen, ibuprofen, and aspirin for pain and inflammation. It's wise to have both adult and child-friendly formulations if you have young ones.
- **Cold Packs/Heat Packs:** Instant cold packs or gel packs for reducing swelling and pain, and heat packs for muscle relief and comfort.

1.3 Infection Prevention

- **Antibiotics:** If possible, and where legal, consider having a stock of common antibiotics like amoxicillin or ciprofloxacin, especially in situations where access to healthcare may be limited.
- **Antiseptic Creams and Ointments:** Neosporin or similar ointments to prevent infection from minor cuts, scrapes, and abrasions.

1.4 Medical Gloves and Masks

- **Nitrile Gloves:** Non-latex gloves to protect yourself and others while administering first aid, especially if dealing with blood or bodily fluids.
- **Face Masks:** Infections can spread quickly, especially in a bug-in scenario. Have disposable face masks on hand to reduce the risk of transmission of airborne diseases.

2. Specialized Tools and Equipment for Advanced Medical Needs

While basic medical kits are great for handling everyday emergencies, when bugging in, it's also crucial to prepare for more severe medical situations. Specialized tools are necessary to provide more advanced care, especially when you're in a prolonged survival situation and may need to perform certain medical procedures yourself.

2.1 Tourniquets

A tourniquet is a critical tool for controlling severe bleeding, particularly in situations where a person has sustained a limb injury with arterial bleeding. The military and emergency medical personnel have long relied on tourniquets to save lives, and they are now recommended for civilian use in first-aid situations.

Recommended: The CAT (Combat Application Tourniquet) or SOF Tactical Tourniquet is reliable, easy to use, and effective in stopping bleeding.

2.2 Splints and Fracture Management

In an emergency where medical help is far away, knowing how to stabilize a broken bone or sprain is essential. A splint is used to immobilize the injured area to prevent further damage and alleviate pain.

Rigid Splints: Use wooden, plastic, or padded splints to stabilize fractures. Pre-made splints are available, but in a survival scenario, any rigid material, such as a stick or rod, can serve as an improvised splint.

Elastic Bandages: These are useful for wrapping and stabilizing injuries like sprains or strains, particularly in the limbs.

2.3 Medical Scissors and Shears

Having a pair of high-quality medical scissors or shears is critical in a survival medical kit. They can be used to cut through clothing, bandages, and even tough materials like seatbelts in the case of an accident.

Recommended: EMT shears are strong and can cut through a variety of materials, making them an indispensable tool in any medical kit.

2.4 Emergency Medical Flashlight

A flashlight designed for medical use is invaluable during an emergency situation, especially in power outages or during nighttime.

Features: Look for a flashlight with a high lumen output, multiple settings, and a red light function (for preserving night vision). Some even come with a built-in stethoscope or thermometer for basic health checks.

2.5 Thermometers

- **Digital Thermometer:** A basic thermometer is a must to monitor for fever, a common symptom in many illnesses.
- **Infrared Thermometer:** For non-contact temperature readings, which can be more convenient and hygienic in certain situations.

2.6 Blood Pressure Cuffs

A blood pressure cuff can be useful for assessing the cardiovascular health of family members or people in your care. If you have someone with a history of hypertension, being able to track their blood pressure during an emergency is crucial.

Automatic Cuffs: These are easier to use and don't require a stethoscope. Make sure to choose one that fits a variety of arm sizes.

3. Medicines and Prescriptions

A critical aspect of medical preparedness is the ability to manage chronic conditions. If you or anyone in your household has pre-existing medical conditions, it's essential to have a sufficient stockpile of necessary medications.

3.1 Stockpiling Prescriptions

Ensure that you have an adequate supply of medications for chronic conditions like diabetes, heart disease, or asthma. Depending on your country's regulations, you may need to work with your healthcare provider to obtain additional refills for your emergency stash.

Considerations: Stockpile medications in their original packaging with clear labels and expiration dates. Always check with a healthcare professional regarding the legality and safety of stockpiling medications.

3.2 Over-the-Counter Medications

In addition to prescription medications, over-the-counter (OTC) remedies are essential for treating common ailments like headaches, allergies, digestive problems, and minor infections. Some common OTC medications to include:

- **Antihistamines:** For allergic reactions, including during seasonal changes.
- **Antidiarrheals:** For cases of food poisoning or stomach upset.
- **Antacids:** For managing heartburn and indigestion.
- **Topical Analgesics:** For managing muscle pain and joint discomfort.

4. Organizing Your Medical Kit

Once you've gathered all the necessary tools and supplies, the next step is organizing your medical kit for easy access during an emergency. Keep in mind the following tips:

- **Use Clear, Labeled Containers:** Store your medical supplies in clear, waterproof containers that are clearly labeled. Ensure you have separate sections for first aid items, medications, and specialized tools.
- **Create a Checklist:** Having a checklist ensures that nothing is overlooked when replenishing or rotating medical supplies. It also helps track expiration dates and conditions of items.
- **Regularly Inspect and Rotate Supplies:** Check for expired medications, dried-out bandages, or broken tools on a regular basis and replace them as necessary.
- **Keep a First-Aid Manual:** Consider including a simple first-aid manual in your kit for easy reference during emergencies.

A medical kit is not just a collection of supplies; it's an essential part of your preparedness strategy. Having a comprehensive medical kit, stocked with both basic and specialized tools, ensures that you can handle medical emergencies, whether they are minor or severe. The right tools, along with the knowledge to use them effectively, can make the difference between life and death in critical situations. By carefully selecting, organizing, and maintaining your medical kit, you'll be ready for whatever challenges arise, allowing you to provide care and support to your loved ones when professional help is out of reach.

7. SCENARIOS AND CRISIS-SPECIFIC PLANNING

Surviving Natural Disasters (Floods, Earthquakes, Storms)

Natural disasters are an unpredictable part of life that can strike without warning, leaving devastation in their wake. Whether you live in an area prone to floods, earthquakes, or severe storms, understanding how to prepare for and survive these events is essential for safeguarding yourself, your family, and your property. Natural disasters often occur with little or no warning, making proactive preparation and quick response key to surviving and minimizing harm.

This chapter will provide you with practical, up-to-date strategies for surviving three of the most common and dangerous types of natural disasters: floods, earthquakes, and storms. Each disaster type requires a specific approach to preparation, response, and recovery, but all share the need for planning, resourcefulness, and resilience. By understanding the science behind these events, knowing how to prepare your home, and learning how to react when disaster strikes, you can increase your chances of survival and protect those you love.

1. Surviving Floods

Flooding is one of the most common and devastating natural disasters. It can occur anywhere, especially in regions with heavy rainfall, melting snow, or near bodies of water. Flooding can damage homes, contaminate water supplies, and cause widespread power outages, making survival difficult if you are unprepared.

1.1 Understanding Flood Risks

Floods can happen quickly, with flash floods occurring within hours of a heavy rainstorm, or more slowly, with river floods developing over days or weeks. The severity of a flood depends on several factors, including:

- **Topography:** Low-lying areas, basements, and floodplains are particularly vulnerable.
- **Weather patterns:** Prolonged rainfall, snowmelt, and hurricanes often cause floods.
- **River and stream overflow:** The overflow of rivers, dams, or levees can result in floodwaters spreading rapidly.

1.2 Preparing for Flooding

- **Know Your Flood Zone:** Check flood maps to determine if your home is located in a flood zone. Understanding local flood risks helps you plan effectively.
- **Elevate Utilities and Appliances:** If you live in a flood-prone area, elevate essential utilities such as electrical panels, heating systems, and appliances above expected flood levels.
- **Flood-Proof Your Home:** Seal windows, doors, and foundations to prevent water from entering your home. Install sump pumps to remove water from your basement and flood barriers around entry points.

- **Emergency Kit:** Your flood preparedness kit should include a battery-powered radio, flashlights, extra batteries, non-perishable food and water, first-aid supplies, medications, important documents, and any necessary evacuation supplies.
- **Insurance:** Flood damage is typically not covered under regular homeowners' insurance. Consider investing in a separate flood insurance policy to protect your property and belongings.

1.3 Responding to Floods

- **Evacuation:** If flooding is imminent or already occurring, evacuate as soon as possible. Follow local authorities' instructions and evacuate to higher ground. Do not wait until floodwaters are rising—early action saves lives.
- **Stay Informed:** Listen to weather reports, flood warnings, and advisories. Utilize weather apps, radios, or NOAA weather channels to stay updated on flood conditions.
- **Don't Drive or Walk Through Flooded Areas:** Even shallow water can be dangerous, as it can quickly sweep away vehicles or people. Avoid traveling in flooded areas at all costs.

2. Surviving Earthquakes

Earthquakes can strike without warning and cause widespread devastation, particularly in regions near fault lines. The ground shaking from an earthquake can result in collapsed buildings, landslides, fires, and tsunamis, which can make survival even more challenging. Earthquakes are unpredictable, but by preparing your home and understanding how to act during and after an earthquake, you can increase your chances of survival.

2.1 Understanding Earthquake Risks

Earthquakes are caused by the sudden release of energy in the Earth's crust, typically along fault lines where tectonic plates meet. The shaking can last anywhere from a few seconds to several minutes, depending on the magnitude of the quake. The impact of an earthquake is measured by magnitude and intensity:

- Magnitude refers to the energy released during the earthquake.
- Intensity measures the effects of the earthquake, including the damage to buildings and infrastructure.

2.2 Preparing for an Earthquake

- **Retrofit Your Home:** If you live in an earthquake-prone area, consider retrofitting your home to make it more resistant to shaking. This might include securing heavy furniture, bracing the foundation, and reinforcing walls and ceilings.
- **Secure Heavy Items:** Ensure that bookshelves, cabinets, mirrors, and other heavy items are secured to walls or stored low to the ground to prevent them from toppling during a quake.
- **Create an Emergency Plan:** Your family should know what to do during and after an earthquake. Designate safe spots, like under a sturdy table or in door frames, and practice "Drop, Cover, and Hold On" drills to ensure everyone knows how to react when the shaking begins.

- **Earthquake Emergency Kit:** This should include water, non-perishable food, first-aid supplies, flashlight, batteries, a whistle, a multi-tool, and a portable charger. If you live in a highly seismic area, consider adding items like a fire extinguisher and sturdy work gloves to your kit.

2.3 Responding to Earthquakes

- **During the Quake:** Drop to the ground, take cover under a sturdy piece of furniture, and hold on until the shaking stops. If you're outdoors, stay in an open area away from buildings, power lines, and trees.
- **After the Quake:** Check for injuries and administer first aid if necessary. Look for hazards such as gas leaks, fires, or structural damage. Avoid using open flames, and do not attempt to move heavy debris unless necessary. Be ready for aftershocks, which can occur after the main quake.
- **Evacuation:** If your building has been damaged, evacuate immediately and move to an open space away from buildings and power lines. Be cautious of the possibility of a tsunami if you live in coastal areas.

3. Surviving Storms (Hurricanes, Tornadoes, and Thunderstorms)

Severe storms, including hurricanes, tornadoes, and thunderstorms, are some of the most destructive natural disasters in terms of both property damage and loss of life. These storms can cause flooding, high winds, lightning, and tornadoes, which can devastate entire communities. Understanding the types of storms and how to prepare for them is critical for survival.

3.1 Understanding Storm Risks

- **Hurricanes:** These powerful tropical storms are characterized by high winds, heavy rainfall, and storm surges. Hurricanes are most common in coastal regions but can impact areas far inland as well.
- **Tornadoes:** These rotating columns of air can form quickly during thunderstorms and cause intense wind damage. Tornadoes often occur in areas known as "Tornado Alley" in the United States.
- **Thunderstorms:** Severe thunderstorms are often accompanied by heavy rain, lightning, hail, and wind, and can lead to flash flooding and power outages.

3.2 Preparing for Storms

- **Know the Warning Signs:** Learn how to recognize storm warnings and watches in your area. This includes listening to weather alerts, watching for changes in the sky (for tornadoes), and knowing evacuation routes.
- **Storm-Proof Your Home:** Install storm shutters or impact-resistant windows, reinforce the roof, and secure loose objects around your home to minimize damage from wind. If you're in a flood-prone area, elevate your home or use flood barriers.
- **Create a Safe Room:** Identify a safe location in your home, such as a basement, storm cellar, or interior room with no windows, where your family can shelter during a storm or tornado.

- **Storm Survival Kit:** This kit should include a battery-powered weather radio, flashlights, extra batteries, food and water for 72 hours, and first-aid supplies. For hurricanes, add waterproof clothing, a personal flotation device, and a power bank for electronics.

3.3 Responding to Storms

During the Storm: Stay indoors and away from windows. If you're in a tornado-prone area, take cover in a basement or storm shelter. In the case of a hurricane or tropical storm, move to higher ground and avoid floodwaters.

After the Storm: Check for any hazards in your environment, such as downed power lines, fallen trees, and flooding. Listen to emergency alerts for recovery instructions, and avoid traveling until it's safe to do so.

Surviving natural disasters requires more than just good luck—it takes preparation, knowledge, and the right tools. By understanding the specific risks posed by floods, earthquakes, and storms, and preparing your home, your family, and your community, you can significantly increase your chances of survival. Stay informed, make proactive choices, and ensure you have the necessary resources to handle any disaster that may come your way. Preparedness is the key to enduring nature's most extreme events, and it's something that can be done, even in the face of unpredictability.

Preparing for Grid-Down and Power Outages

Power outages are one of the most common and disruptive types of emergency events. Whether due to natural disasters like hurricanes, snowstorms, wildfires, or man-made factors such as cyberattacks or grid failures, being prepared for a grid-down scenario is crucial for maintaining safety and comfort during extended power outages. As more of our daily lives depend on electricity—whether for cooking, heating, refrigeration, or communication—the ability to survive and function without power can be the difference between inconvenience and crisis. Preparing for grid-down situations not only involves having the necessary backup systems in place but also understanding the risks and challenges associated with long-term power loss.

In this section, we will explore the essential steps for preparing for power outages, covering everything from basic emergency kits to long-term power solutions, along with the safety protocols you should follow to protect yourself and your family when the grid goes down.

1. Understanding Grid-Down Risks

A grid-down event refers to the loss of electrical power from the main utility grid, which supplies power to homes, businesses, and essential infrastructure. Power outages can be caused by a variety of factors, including:

- **Severe Weather Events:** Hurricanes, tornadoes, snowstorms, and flooding can down power lines and cause widespread outages.
- **Natural Disasters:** Earthquakes, wildfires, and even solar flares can disrupt electrical grids.
- **Cyberattacks or Terrorism:** As reliance on digital infrastructure grows, power grids are increasingly vulnerable to attacks that could cause prolonged outages.

- **Aging Infrastructure:** Many power grids worldwide are aging and increasingly unreliable, especially in regions with outdated systems that have not been maintained or upgraded.
- **Grid Overload:** Extreme temperatures, such as heat waves or cold snaps, can lead to increased demand on the grid, potentially causing failures due to overload.

The potential for prolonged outages in these scenarios makes preparation essential. Even short-term power outages can disrupt daily activities, spoil food, and jeopardize health and safety.

2. Preparing for Short-Term Power Outages

While power outages lasting just a few hours or a day are relatively common, they still require preparedness to minimize their impact. Here's how to prepare for short-term power outages:

2.1 Emergency Kits for Power Outages

Having an emergency kit is one of the first steps in preparing for any type of grid-down situation. This kit should be readily accessible and contain the essentials to help you manage during a temporary outage:

- **Flashlights:** Always have battery-operated or hand-crank flashlights to avoid being left in the dark. Stock extra batteries for these.
- **Batteries:** A wide range of batteries, including for portable devices, radios, and flashlights.
- **Portable Power Bank:** Power banks or solar chargers are essential for charging phones and communication devices without access to electricity.
- **Battery-Powered Radio:** Stay informed during an outage by having a battery-powered or hand-crank radio that can pick up emergency broadcasts and weather updates.
- **Water and Non-Perishable Food:** At least a 72-hour supply of water and ready-to-eat meals (canned foods, protein bars, etc.). Don't forget a manual can opener if you have canned goods.
- **First-Aid Kit:** A well-stocked first-aid kit can be vital in emergency situations.
- **Blankets or Sleeping Bags:** If the outage occurs in cold weather, these can help keep you warm.
- **Medications:** Keep extra medications and copies of prescriptions.

2.2 Refrigerator and Food Safety

One of the most immediate concerns during a short-term power outage is food safety. The temperature inside your fridge can rise quickly after the power goes out, leading to spoiled food. Here's how to manage:

- **Keep the Fridge Door Closed:** The less you open the fridge, the longer the food inside will stay cold. A full refrigerator will stay cold for about 4 hours, while a freezer will keep food frozen for 24 to 48 hours if unopened.
- **Coolers and Ice:** Keep a supply of coolers and ice on hand to store perishable items if the outage is expected to last more than a few hours. This is particularly important during summer months when temperatures can rise quickly.

2.3 Safe Heating and Cooling Options

Without electricity, heating and cooling systems may not function, especially during extreme temperatures. Prepare by having:

- **Battery-Powered Fans:** Small fans powered by batteries or solar panels can help circulate air during hot weather.
- **Portable Heaters:** Battery-powered or propane-based space heaters can provide warmth in cold conditions.
- **Extra Warm Clothing and Bedding:** Stockpile warm clothes, hats, gloves, and blankets to keep you and your family comfortable in colder weather.

3. Preparing for Long-Term Power Outages

While short-term outages are more common, it's important to consider longer, more severe power loss situations. These might arise from major storms, infrastructure breakdowns, or even national-scale grid failures. Here are strategies for preparing for a long-term grid-down scenario:

3.1 Alternative Power Sources

In a long-term outage, having a backup power source is essential. The two most common solutions are generators and renewable energy systems.

- **Generators:** Backup generators, either portable or standby, are one of the most reliable ways to keep your lights on and appliances running when the power is out. When purchasing a generator, consider factors such as:
- **Fuel Type:** Gasoline, propane, and diesel are common fuel sources, with propane often being the most long-lasting and safe option.
- **Power Output:** Choose a generator with enough wattage to power your essential devices (refrigerators, medical equipment, lights, etc.).
- **Maintenance:** Generators require regular maintenance, including oil changes and fuel stabilizer to ensure they're ready when needed.
- **Carbon Monoxide Safety:** Always operate generators outdoors and away from windows, vents, or doors to prevent the build-up of carbon monoxide.
- **Solar Power:** Solar power can be a more sustainable and longer-lasting solution. Solar panels can be used to generate electricity during the day, and with a good battery storage system, you can store excess power for use at night or during cloudy days. This solution can be expensive upfront but will pay off in the long run, especially for off-grid living.
- **Wind and Water Power:** Small-scale wind turbines or hydroelectric generators (if you live near a water source) can also be part of your backup plan. These systems are ideal for long-term, off-grid scenarios but require significant space and setup.

3.2 Home Energy Conservation

Conserving energy during a grid-down situation is key to prolonging your supply of backup power. Some tips include:

- **Unplug Non-Essential Electronics:** Electronics that are plugged in but not in use can drain power. Unplugging these devices when not in use helps preserve energy for essential devices.
- **Use Energy-Efficient Lighting:** LED lights consume far less power than traditional incandescent bulbs, and they can last much longer. Having a supply of LED lanterns or rechargeable lamps is useful during outages.
- **Seal Drafts:** If you rely on a generator or alternative heat source, reducing heat loss by sealing windows and doors can help maintain warmth inside the home.

3.3 Water Supply

Without electricity, municipal water supplies may be cut off, especially if the outage is widespread or prolonged. Consider these options for securing an alternative water supply:

- **Rainwater Harvesting:** Set up rainwater collection systems to collect and filter rainwater for drinking, cooking, and sanitation purposes.
- **Water Filtration Systems:** In addition to storing water in bottles, use water filters like reverse osmosis systems or portable filters to ensure a safe and sustainable water supply.
- **Water Storage Containers:** Invest in water storage containers or large tanks to keep a reserve of water on hand. Store at least one gallon of water per person per day for drinking, cooking, and hygiene.

3.4 Communication Systems

Without electricity, most traditional communication systems, such as landlines and cell phones, may not function. Here's how to stay connected:

- **Battery-Powered or Hand-Crank Radio:** Stay informed with a hand-crank or battery-powered radio that can pick up emergency alerts and weather information.
- **Satellite Phones:** For long-term preparedness, satellite phones or radios that don't rely on the grid can provide a vital connection if phone lines go down.
- **Local Community Networks:** Build relationships with neighbors and local communities who are also preparing for emergencies. A local network can provide support, share resources, and act as a communication link during a crisis.

Preparing for grid-down and power outage scenarios involves taking a holistic approach—planning ahead, securing the necessary resources, and ensuring that both your home and your family are ready for any disruption in power. By understanding the risks and having the right backup systems in place, you can face power outages with confidence and keep your family safe and secure. With a little foresight and resourcefulness, you can not only survive but thrive during grid-down emergencies, and with this knowledge, you will be prepared for whatever the future holds.

Handling Pandemics and Biohazards

In the modern world, pandemics and biohazards present significant and evolving challenges for individuals, families, and communities. A pandemic is a global outbreak of a contagious disease that affects large numbers of people across multiple countries or continents. Biohazards, on the other hand, refer to biological substances—such as viruses, bacteria, fungi, and toxins—that pose a threat to human health and the environment. Both pandemics and biohazards can disrupt daily life, put strain on health systems, and challenge the infrastructure of even the most developed nations. The COVID-19 pandemic of 2020 served as a stark reminder of how quickly diseases can spread and how crucial it is to be prepared.

This chapter provides a comprehensive guide to handling pandemics and biohazards, offering strategies for preparedness, protection, and response. The key to managing these emergencies lies in proactive planning, understanding the risks, and adapting to new information as situations evolve. From maintaining hygiene to preparing for quarantine, and from understanding symptoms to knowing how to protect your family, this section will guide you through the critical aspects of surviving during a pandemic or biohazard event.

1. Understanding Pandemics and Biohazards

1.1 What is a Pandemic?

A pandemic is typically defined as an outbreak of a disease that spreads across countries or continents, affecting a large portion of the population. Unlike localized outbreaks or epidemics, pandemics are widespread and can overwhelm public health systems, creating both health and societal crises. The COVID-19 pandemic, caused by the SARS-CoV-2 virus, is one of the most significant recent examples, affecting nearly every country worldwide and resulting in widespread illness, death, and economic disruption.

Pandemics are usually caused by infectious agents such as viruses or bacteria that are easily transmitted from person to person. The severity of the pandemic can vary depending on factors such as:

The nature of the pathogen (e.g., how contagious it is, how easily it spreads, and how deadly it is)

Global travel patterns (international travel facilitates the rapid spread of diseases)

Population density (urban areas with high populations can exacerbate the spread)

Healthcare system preparedness (the ability of hospitals and public health institutions to respond to a surge in cases)

1.2 What are Biohazards?

Biohazards refer to any biological agent—whether bacteria, viruses, fungi, or toxins—that can harm human health or the environment. Biohazards can be naturally occurring or man-made. Examples of biohazards include:

Infectious diseases: Bacteria or viruses that can cause diseases like influenza, anthrax, or HIV/AIDS.

Bioterrorism agents: Pathogens intentionally used to cause harm or fear, such as smallpox, plague, or ricin.

Toxins: Harmful substances like botulinum toxin, which can be fatal in very small amounts.

Contaminated food or water: Foodborne illnesses caused by pathogens like E. coli or Salmonella.

Biohazardous events can range from small-scale localized incidents (such as foodborne outbreaks) to large-scale, potentially catastrophic incidents (like bioterrorism attacks).

2. Pandemic and Biohazard Preparedness

2.1 Building a Pandemic-Ready Emergency Kit

Preparing an emergency kit specifically for a pandemic or biohazard event is essential. Your kit should contain items that protect you from exposure to the disease and enable you to function at home for an extended period. Items to include:

- **Personal Protective Equipment (PPE):** Masks (N95 or similar), gloves, face shields, and goggles to protect yourself from airborne viruses or bacteria.
- **Sanitation Supplies:** Hand sanitizers, disinfectant wipes, bleach, and disinfectant sprays to clean surfaces and kill pathogens. Alcohol-based sanitizers should contain at least 60% alcohol to be effective.
- **Medications:** Over-the-counter medications such as pain relievers, fever reducers, and cough medicines. Ensure you have a 30-day supply of any prescription medications you take regularly.
- **Non-perishable Food:** Stockpile shelf-stable foods that are easy to prepare, such as canned goods, dried beans, rice, pasta, and energy bars.
- **Water:** At least one gallon of water per person per day, for drinking and sanitation.
- **Hygiene Products:** Soap, toilet paper, feminine hygiene products, and baby wipes.
- **First-Aid Kit:** Bandages, antiseptics, thermometers, and other medical supplies to treat minor injuries or illnesses at home.
- **Cleaning Supplies:** Large garbage bags, trash bins, and protective bags for safely disposing of contaminated waste.

This kit should be easily accessible, and you should be prepared to shelter in place for an extended period, avoiding public places to minimize exposure to pathogens.

2.2 Staying Informed During a Pandemic or Biohazard Event

Staying informed during a pandemic or biohazard event is essential for making decisions about your health and safety. Here are the best ways to stay updated:

Official Sources: Follow information from trusted health authorities like the Centers for Disease Control and Prevention (CDC), the World Health Organization (WHO), and local public health departments for the latest updates and guidelines.

Reliable News Outlets: Be cautious of misinformation during pandemics. Stick to well-established news sources that cite credible experts and avoid panic-driven rumors.

Emergency Alerts: Sign up for local emergency alerts to receive notifications about important events, quarantine orders, or other updates.

2.3 Quarantine and Isolation Strategies

In the event of a contagious disease outbreak, quarantine and isolation measures may be necessary to prevent the spread of the disease. These strategies involve staying at home and avoiding contact with others, even if you feel healthy. To prepare:

Create a Quarantine Space: Designate a separate area of the house for anyone who becomes ill or has been exposed to a contagious disease. This space should include basic necessities such as food, water, and personal hygiene items.

Limit Contact: Avoid direct contact with anyone who is ill or who has potentially been exposed to the disease. If possible, designate one caregiver to tend to the sick while minimizing their exposure to the rest of the household.

Self-Monitoring: Keep track of symptoms and monitor the health of everyone in your household. Use thermometers, pulse oximeters, and other tools to check for fever, oxygen levels, or other health indicators.

Isolation Supplies: Ensure the sick person has access to their own set of hygiene products, food, and water. This limits the spread of pathogens to the rest of the household.

3. Protecting Your Home and Community

3.1 Biohazard Containment and Cleanup

In the event of a biohazard exposure, it is important to contain and decontaminate the affected area to prevent further exposure. Consider the following:

- **Containment:** Seal off the affected area to limit the spread of biological agents. Use plastic sheeting, tape, or makeshift barriers if necessary to create a containment zone.
- **Decontamination:** Disinfect surfaces thoroughly, especially high-touch areas like doorknobs, light switches, and countertops. Use EPA-approved disinfectants that are effective against viruses and bacteria.
- **Waste Disposal:** Properly dispose of contaminated materials, such as used tissues, masks, gloves, or cleaning supplies. Use sealed bags to dispose of these items safely.

3.2 Community Preparedness

During a pandemic or biohazard event, community-level preparedness can make a significant difference. Community actions include:

- **Community Response Teams:** Organize neighborhood groups to help deliver food, medications, or essential supplies to those who are quarantined or at risk.

- **Mutual Aid Networks:** Share resources with neighbors, such as medical supplies, food, or personal protective equipment. This helps ensure that no one is left behind during the crisis.
- **Communication:** Establish local communication channels (e.g., social media groups, text alerts, or community apps) to keep everyone informed about safety protocols, available resources, and changes in the situation.

Pandemics and biohazardous events present complex and unique challenges, but with the right knowledge, preparation, and mindset, you can protect yourself, your family, and your community. By understanding the risks, building a robust emergency kit, staying informed, and employing isolation and containment strategies, you'll be well-equipped to manage a pandemic or biohazard event. It is critical to remain vigilant, adapt to new information, and prioritize health and safety above all. Preparing for these events is not just about survival—it's about ensuring you have the resilience and resources to navigate through the worst of times and emerge stronger.

Protecting Your Home During Civil Unrest

Civil unrest is an unfortunate reality in today's world. Protests, riots, demonstrations, or even armed conflicts can erupt for a variety of reasons, ranging from political disagreements to social issues. While some unrest may be localized, others can spread rapidly and create dangerous situations for individuals and families. It is essential to understand how to protect your home and loved ones during such events to ensure your safety, prevent property damage, and maintain a level of security until the situation stabilizes.

Civil unrest can take many forms, including peaceful protests that turn violent, large-scale demonstrations, and riots that escalate into widespread chaos. These events can cause significant disruptions to daily life, affect transportation, impact the economy, and put public services at risk. When civil unrest occurs, your primary goal should be to ensure your safety, the safety of your family, and the protection of your home.

1. Understanding the Risks of Civil Unrest

Civil unrest often results in violent clashes between demonstrators and law enforcement or between opposing groups. These confrontations can escalate quickly, causing widespread destruction, fires, looting, and even armed encounters. Some of the risks associated with civil unrest include:

- **Violence and Vandalism:** Riots and protests that turn violent can lead to property damage, looting, and even personal injury. Windows may be shattered, cars set on fire, and buildings vandalized.
- **Civil Disobedience:** Strikes, barricades, and other forms of civil disobedience can disrupt the flow of goods and services, prevent access to key infrastructure, and make it difficult to leave or enter your home.
- **Terrorism and Attacks:** During times of civil unrest, there is a higher risk of extremist groups taking advantage of the chaos to carry out attacks, which could target public infrastructure, government buildings, or even private homes.

- **Political Polarization:** As civil unrest is often politically motivated, tensions between opposing groups can increase, putting you at risk if you are caught in the middle.

Understanding the potential dangers of civil unrest will allow you to better prepare your home and develop strategies to minimize the risks.

2. Preparing Your Home for Civil Unrest

Preparation is key when it comes to protecting your home during civil unrest. There are several steps you can take to fortify your property, reduce your vulnerability, and increase your chances of staying safe:

2.1 Secure Doors and Windows

One of the most important aspects of home security during civil unrest is reinforcing entry points. Rioters or individuals looking to loot may target homes that appear unprotected. Reinforcing doors and windows can make your home a less attractive target.

- **Reinforce Doors:** Install solid-core or metal doors, which are harder to break through than hollow wood doors. Use deadbolt locks and additional security bars if necessary. Consider installing security doors with peepholes or cameras so you can monitor activity outside without exposing yourself.
- **Reinforce Windows:** Consider using security film on windows to prevent glass from shattering upon impact. Install window bars or use plywood boards to cover windows and prevent them from being broken. If you're not home during the unrest, this can provide an additional layer of protection.
- **Sliding Doors and Windows:** Install a security bar on sliding doors or windows, which are often weak entry points. You can also place a metal rod or piece of wood in the track to prevent the door from being forced open.

2.2 Create Perimeter Defenses

Fortifying the perimeter of your property is critical during civil unrest. If the unrest spreads to your neighborhood, your home will be better protected from both looters and potential violence.

Fencing: A solid fence, particularly one made of metal or wood, can deter individuals from attempting to enter your property. Make sure gates are secure with high-quality locks and that the fence is high enough to prevent easy access.

Lighting: Adequate lighting is crucial for deterring criminal activity. Motion-sensor lights around your property will alert you to any movement and can make intruders think twice before approaching.

Landscaping: Use landscaping strategically to make your property less appealing and harder to navigate. Thorny bushes or plants can create natural barriers that deter trespassers. Avoid using overly dense vegetation near doors and windows that could provide cover for someone attempting to break in.

2.3 Emergency Evacuation Plans

In extreme cases, when civil unrest becomes violent or widespread, you may need to evacuate your home quickly. Establishing an evacuation plan and ensuring your family knows how to react can significantly reduce the risks.

- **Evacuation Routes:** Plan multiple routes to leave your home. Know your neighborhood and identify safe zones (such as parks or open areas) where you can seek shelter. Avoid main roads, as these may be blocked by barricades, traffic jams, or violent crowds.
- **Meeting Points:** Choose a safe meeting place for family members if you become separated during an evacuation. This could be a nearby relative's home or a designated public location.
- **Bug-Out Bags:** Have bug-out bags ready for each family member, containing essential items like water, food, medical supplies, flashlights, and any necessary documents. Ensure that these bags are stored in a safe, easily accessible location.

3. Communication During Civil Unrest

Communication is a key component of staying safe during civil unrest. Without reliable communication, you may find it difficult to stay updated on the evolving situation and communicate with family members or neighbors.

- **Smartphones and Apps:** Use emergency communication apps that allow you to receive real-time updates on local conditions. Apps like FEMA, Red Cross, or local alert systems can help you stay informed about curfews, evacuation orders, and emergencies in your area.
- **Two-Way Radios:** In case cell networks become overloaded or disrupted, two-way radios (walkie-talkies) can be an effective communication tool. Ensure all family members are trained on how to use them.
- **Social Media:** Follow local authorities, news outlets, and reliable sources on social media platforms for live updates about the situation. Social media can also be a way to share information with neighbors, but always verify the source of information to avoid spreading rumors.

4. Defending Your Home if Necessary

In some situations, the threat to your home may become direct. If your home is under threat from intruders, looters, or vandals, having a plan to defend it may be necessary.

- **Self-Defense Weapons:** Depending on your local laws, you may want to consider acquiring self-defense weapons such as firearms, pepper spray, or a baton. It is critical that everyone in your home understands how to safely use any weapons and is trained in their proper handling.
- **Safe Room:** Create a safe room within your home that is fortified and has enough supplies to withstand a prolonged stay. This could be a basement or an interior room without windows. The safe room should contain enough food, water, and medical supplies to last for several days, as well as a phone or means of communication.

- **Non-Violent Defense**: Sometimes, it's best to avoid confrontation. If the unrest escalates and looting or violence is imminent, consider hiding valuables, turning off lights, and staying out of sight until the threat passes.

5. Staying Informed During Civil Unrest

During civil unrest, situations can evolve quickly, and staying informed is essential for making the right decisions.

Local News and Radio: Tune in to local news channels or radio stations for updates. If the internet is down, having a battery-powered radio can be invaluable for receiving emergency broadcasts.

Monitor Social Media: Social media platforms can provide live updates from local authorities, news outlets, and individuals on the ground. However, be cautious of misinformation—always cross-check information with reliable sources.

Civil unrest is unpredictable, and the best way to protect your home and family is through preparation and awareness. By reinforcing your home's entry points, developing an emergency evacuation plan, staying informed, and securing your perimeter, you can dramatically increase your chances of protecting your property during an episode of civil unrest. At all times, remember that your safety is the top priority, and avoiding confrontation or exposure to violence is often the best course of action.

8. SELF-SUFFICIENCY AT HOME

Setting Up Off-Grid Energy Solutions

In the context of survival and long-term preparedness, having an off-grid energy solution is critical. The ability to power your home without relying on traditional electricity grids is not only a means of maintaining comfort but can also be a matter of life and death in times of crisis. Whether you're dealing with power outages caused by storms, natural disasters, civil unrest, or even broader grid failures, setting up an off-grid energy solution can provide you with the independence and peace of mind needed to navigate emergencies with confidence.

In this chapter, we'll explore the various off-grid energy options available, how to set them up, and the advantages and challenges of going off-grid. We'll focus on practical, cost-effective strategies that you can implement in your home to ensure energy independence during crises, as well as sustainable solutions for long-term use.

1. Understanding Off-Grid Energy Solutions

Off-grid energy solutions are systems that allow you to generate, store, and use electricity without being connected to the public utility grid. These systems are typically composed of a combination of power generation methods (solar, wind, hydro), energy storage systems (batteries), and backup generators. Off-grid systems can be designed to provide power to individual households, small communities, or even entire farms, depending on the scale and energy needs.

While many homeowners are familiar with traditional grid-connected electricity, off-grid systems provide a level of resilience and autonomy that is invaluable, especially when the grid is unavailable, overloaded, or vulnerable to attacks.

Key components of off-grid energy systems include:

- **Energy Generation:** The most common sources of off-grid electricity generation are solar panels, wind turbines, and micro-hydro systems.
- **Energy Storage:** Batteries are used to store energy for use when the generation source isn't producing power (e.g., at night or during calm winds).
- **Power Management:** This includes inverters (which convert DC to AC electricity) and charge controllers (which regulate battery charging) to ensure optimal energy use.

Each of these components plays a role in creating a system that can meet your energy needs while remaining independent of the grid.

2. Key Off-Grid Energy Options

2.1 Solar Power

Solar power is the most popular and widely used form of off-grid energy generation. Solar photovoltaic (PV) panels convert sunlight directly into electricity, providing a clean and renewable source of power.

Solar is highly scalable, meaning that you can design a system that fits your specific energy requirements. Here's how you can set up solar power for off-grid use:

- **Solar Panels:** These are the heart of the system and collect sunlight to generate electricity. Panels come in various sizes and configurations, and their efficiency depends on factors like sunlight hours, orientation, and tilt.
- **Inverters:** Solar panels typically generate direct current (DC) power, but most appliances use alternating current (AC) power. An inverter is necessary to convert DC to AC so that your household devices can run.
- **Batteries:** To ensure you have power when the sun isn't shining, you'll need a battery bank to store energy. Lithium-ion batteries are popular for their long lifespan and efficiency, but lead-acid batteries are a more affordable option.
- **Charge Controller:** This device regulates the flow of energy from the panels to the batteries, ensuring that the batteries aren't overcharged or discharged too much, which can shorten their lifespan.

Advantages of Solar Power:

- Abundant, renewable resource.
- Low maintenance requirements.
- Scalable for small or large homes.

Challenges:

- Upfront cost for purchasing and installing the system.
- Requires ample sunlight for maximum efficiency.

2.2 Wind Power

Wind turbines convert the kinetic energy from wind into electricity. This option is ideal if you live in an area with consistent wind speeds. Wind energy systems can either be grid-tied or off-grid, and they come in various sizes, ranging from small, household turbines to large commercial models.

- **Small-Scale Wind Turbines:** These turbines are typically used to generate electricity for homes or small communities. They can be installed on rooftops or on free-standing poles in areas with sufficient wind.
- **Battery Storage:** Like solar power, wind energy requires battery storage to store the energy produced during windy periods for use when wind speeds are low or during the night.

Advantages of Wind Power:

- Ideal for locations with consistent and strong winds.
- Can complement solar power by providing energy at night or during cloudy periods.

Challenges:

- Wind speeds must be sufficient for energy production.

- Maintenance requirements for turbines.
- Requires adequate space for installation.

2.3 Micro-Hydro Power

Micro-hydro systems generate electricity by harnessing the power of flowing water, typically from streams or rivers. These systems are often used in remote locations where other forms of renewable energy are impractical.

Water Flow: Micro-hydro systems require a consistent water flow, and the amount of power generated is dependent on the flow rate and the height (head) of the water.

Turbines: Water turbines generate electricity by turning mechanical energy from flowing water into electrical energy, similar to a wind turbine.

Storage: Like wind and solar, hydro systems also require batteries to store excess power.

Advantages of Micro-Hydro Power:

- Provides a consistent, reliable energy source if you have access to a water source.
- Low operating costs once installed.

Challenges:

- Installation can be expensive and requires specific geographic conditions (a steady flow of water).
- Environmental regulations may restrict installation in certain areas.

3. Hybrid Systems: Combining Energy Sources

While each of the individual energy sources mentioned above can be effective on their own, combining multiple sources of energy in a hybrid system can provide more reliable, consistent power. Hybrid systems often combine solar, wind, and/or micro-hydro with a battery bank to ensure that power is available no matter the time of day, weather conditions, or season.

For example, a typical off-grid hybrid setup might include:

- Solar panels for daytime power generation.
- Wind turbines to generate power during windy conditions, especially at night when solar isn't producing.
- Batteries to store excess energy for later use.
- Backup generator to provide power during prolonged periods of low sunlight or wind.

Advantages of Hybrid Systems:

- Maximizes energy production by utilizing multiple sources.
- Provides more stability and reliability than relying on a single energy source.
- Reduces dependence on backup generators.

Challenges:

- Higher initial installation cost.
- Requires careful planning to integrate the different energy sources effectively.

4. Backup Generators

While renewable energy solutions such as solar, wind, and micro-hydro can provide significant power independence, they often need to be supplemented with backup power sources during extended periods of low energy production. A generator provides a reliable solution in such cases, offering both short-term and long-term backup power.

Generators come in several types:

- **Gasoline or Diesel Generators:** These are typically used as a short-term solution to power essential systems during an emergency.
- **Propane Generators:** These are more fuel-efficient and provide an alternative to gasoline-powered generators.
- **Dual-Fuel Generators:** Some generators can run on both gasoline and propane, providing greater flexibility and ensuring that you have a fuel source available no matter the situation.

Advantages of Backup Generators:

- Provides an immediate and reliable source of energy when renewable sources are insufficient.
- Highly portable and can be used for multiple purposes (powering appliances, charging batteries, etc.).

Challenges:

- Requires fuel, which may not be readily available during prolonged emergencies.
- Maintenance is needed to ensure the generator works efficiently.

5. Maintenance and Longevity of Off-Grid Systems

One of the critical factors in setting up off-grid energy solutions is ensuring that your system remains functional and reliable in the long run. Regular maintenance is essential to avoid costly repairs and ensure that your system continues to meet your energy needs.

Solar Panels: Clean the panels regularly to remove dirt, dust, and debris that may block sunlight. Periodically inspect the wiring and connections for wear and tear.

Wind Turbines: Inspect the turbine blades, bearings, and electrical components to ensure they are operating smoothly. Check for corrosion, particularly in coastal areas where saltwater may be an issue.

Batteries: Ensure that batteries are maintained properly, including checking the water levels in lead-acid batteries and monitoring the charge levels of lithium-ion batteries. Replace batteries when their capacity starts to degrade.

Backup Generators: Perform regular testing and maintenance, including changing the oil, cleaning filters, and ensuring that the generator runs properly.

Setting up an off-grid energy system requires careful planning, but the benefits far outweigh the initial investment. Whether you are looking to be more self-reliant, reduce your carbon footprint, or prepare for emergencies, off-grid energy solutions provide a reliable and sustainable way to power your home during times of need. By choosing the right combination of energy sources, properly maintaining your system, and understanding how to integrate these solutions into your daily life, you can gain independence and resilience in the face of challenges.

Water Harvesting and Filtration Systems

In an emergency situation, access to clean water is one of the most critical concerns. According to the World Health Organization (WHO), water scarcity affects over two billion people globally, and the availability of clean water can be drastically reduced during natural disasters, civil unrest, or other crises. This makes water harvesting and filtration systems an essential part of any comprehensive off-grid survival plan.

Water harvesting and filtration are not just for those living in arid regions or remote locations; they are practical, lifesaving solutions for everyone preparing for any type of emergency, from power outages to pandemics. Being able to capture and purify water will give you autonomy and peace of mind, knowing you can provide clean water to your family during times when municipal supplies may be compromised or inaccessible.

In this chapter, we will explore the most effective and sustainable water harvesting methods and filtration systems. We will look at techniques ranging from rainwater harvesting to advanced filtration technologies and how you can implement them to secure an adequate water supply for your home.

1. Understanding Water Harvesting

Water harvesting refers to the collection and storage of rainwater for use in drinking, irrigation, or other domestic purposes. It is a simple, sustainable practice that can be done on any scale, from a small household system to a large farm operation.

Key Methods of Water Harvesting:

Rainwater Harvesting: This is the most common method of water collection and involves capturing rainfall from rooftops or other surfaces and storing it in barrels, tanks, or cisterns. The collection system generally includes gutters, downspouts, filters, and storage containers. The amount of water you can collect depends on the size of your catchment area (i.e., your roof), the amount of rainfall, and the capacity of your storage system.

Surface Runoff Harvesting: This method involves capturing rainwater that runs off impervious surfaces like roads, driveways, or fields. Specialized systems, such as swales, dams, and ponds, are used to capture and store this runoff. It is more complicated than simple rainwater harvesting but may be a viable option in areas with heavy rainfall or a significant amount of runoff.

Groundwater Recharge: While not a collection method per se, groundwater recharge involves using harvested water to replenish underground aquifers through infiltration techniques. In areas with shallow groundwater, this method can help replenish natural water reserves for long-term sustainability.

2. Components of a Water Harvesting System

A water harvesting system typically consists of four key components:

- **Catchment Area:** This is the surface from which rainwater is collected. In most cases, the catchment area will be your roof, but it can also include other surfaces like paved areas or dedicated rain gardens.
- **Conveyance System:** This system directs the rainwater from the catchment area to the storage containers. It includes gutters, downspouts, and pipes. Proper slope and sizing are critical to efficiently carry water from the catchment area to storage.
- **Filtration System:** Rainwater may contain debris, dust, leaves, and pollutants, so it is essential to filter out these contaminants before storage and use. Basic filtration can be done with mesh screens, while more advanced systems use filters or first-flush diverters.
- **Storage:** The water collected is stored in containers such as rain barrels, tanks, or underground cisterns. These should be made of non-toxic, UV-resistant materials to prevent contamination and algae growth. The storage size depends on your family's water needs and your local rainfall patterns.

3. Advanced Water Filtration Techniques

Once rainwater is harvested, it often needs to be filtered to ensure it is safe for consumption and use. Several filtration techniques can help remove harmful pathogens, chemicals, and particulates. The method you choose will depend on your water quality, your filtration needs, and your available resources.

3.1 Mechanical Filters

These are the most basic type of filtration and include mesh screens, cloth filters, and sand filters. Mechanical filters remove large particles like leaves, dirt, and debris. However, they are not enough to remove pathogens or dissolved chemicals.

- **Mesh Screens:** These are used at the entry point of your system (such as in the gutters or at the downspouts) to filter out leaves and large debris. Fine mesh filters can remove small particles.
- **Sand Filters:** Sand filtration is a simple, low-cost method where water is passed through layers of sand to remove suspended solids. It's effective at removing particulate matter but doesn't address bacteria or viruses.

3.2 Activated Carbon Filters

Activated carbon (or activated charcoal) is one of the most effective filtration media for removing chlorine, heavy metals, and organic compounds from water. It works through adsorption, where contaminants stick to the surface of the carbon. Activated carbon filters can be purchased as standalone units or incorporated into larger filtration systems.

Advantages: Activated carbon is effective at improving taste and removing a wide range of chemicals, including pesticides, herbicides, and volatile organic compounds (VOCs).

Disadvantages: It doesn't remove all contaminants, such as bacteria, viruses, and salts.

3.3 Reverse Osmosis (RO) Systems

Reverse osmosis is a more advanced filtration technique that uses a semipermeable membrane to remove a wide range of contaminants, including bacteria, viruses, salts, and heavy metals. Water is forced through the membrane, leaving contaminants behind.

Advantages: RO systems are highly effective at purifying water and can remove nearly all harmful substances.

Disadvantages: They are energy-intensive, produce wastewater, and require periodic maintenance. They are also slow and may require high-pressure pumps to work effectively.

3.4 UV Purification

Ultraviolet (UV) light is an effective method for killing bacteria, viruses, and other microorganisms. UV purification doesn't remove particulates or chemicals, but it is a highly effective method for sterilizing water that has been pre-filtered. UV purifiers can be standalone devices or integrated into larger water treatment systems.

Advantages: UV purifiers are chemical-free and fast, sterilizing water without affecting its taste or quality.

Disadvantages: They require electricity to operate and may not be as effective if water contains a high level of sediment or organic material.

3.5 Ceramic Filters

Ceramic filters are made from porous ceramic material that physically blocks contaminants. They are effective at removing bacteria, protozoa, and large particles from water, though they don't address viruses or dissolved chemicals.

Advantages: Low maintenance, chemical-free, and affordable.

Disadvantages: Limited in their ability to remove certain pathogens and chemicals.

4. Setting Up a Water Filtration System

To effectively use harvested rainwater, you need a robust filtration system that can purify water to a standard that's safe for drinking, cooking, and hygiene. The process typically includes the following steps:

- **Install a First Flush Diverter:** A first flush diverter ensures that the first flow of water, which is most likely to be contaminated with dust, bird droppings, and other debris, is discarded before it enters the storage system.

- **Pre-Filter Water:** Use a mesh screen or pre-filtration system to remove leaves and large debris before the water enters the main filtration unit.
- **Choose Appropriate Filtration:** Depending on your needs, install activated carbon filters, ceramic filters, or even a reverse osmosis unit for advanced purification. In a survival situation, you may need to prioritize compact, multi-stage filters that can handle a variety of contaminants.
- **Post-Storage Filtration:** After storing the water, you may wish to install an additional filtration system for daily use, such as a UV sterilizer or a final-stage activated carbon filter for taste and chemical removal.

Water harvesting and filtration are essential survival skills that provide an affordable, sustainable solution for water security. Whether you're relying on rainwater or other sources, having an efficient filtration system ensures that the water you collect is clean, safe, and usable for drinking, cooking, and sanitation. By understanding the various harvesting methods, filtration techniques, and setup strategies available, you can establish a reliable and independent water supply for you and your family, ensuring survival in any emergency situation. With these systems in place, you gain autonomy and peace of mind, knowing that you can weather any storm, literally and figuratively, with access to clean, fresh water.

Indoor and Outdoor Gardening for Sustenance

In any survival or preparedness scenario, having a reliable source of food is a critical factor. Whether you are preparing for a natural disaster, a civil unrest situation, or simply aiming to be more self-sufficient, gardening provides a sustainable, low-cost way to produce fresh food. Both indoor and outdoor gardening have their unique advantages and challenges, but when done effectively, they can ensure you have access to nutritious food during times of crisis.

This chapter explores how to use gardening to sustain yourself and your family, whether through growing food in your backyard or on windowsills, countertops, or balconies. We will cover the basic principles of gardening, the types of food that are easy to grow in both indoor and outdoor environments, and the tools and techniques you need to ensure your garden thrives during challenging circumstances.

1. The Importance of Self-Sufficient Gardening in Crisis Situations

In emergencies, food supply chains are often disrupted, leaving communities without access to fresh produce and essential staples. For example, during natural disasters like hurricanes or floods, or man-made disruptions like political instability, food deliveries may be delayed or completely cut off. In these situations, having a sustainable food source that you can rely on in your own home or immediate surroundings is invaluable.

Gardening for sustenance can be broken down into two categories:

Indoor Gardening: A method of growing food inside your home, ideal for those in urban areas or those without access to a large outdoor space.

Outdoor Gardening: Growing food in your yard, garden, or on a larger piece of land. This method is perfect for those with access to more outdoor space and allows for a greater variety of crops.

2. Indoor Gardening for Food Security

Indoor gardening has become increasingly popular as people seek ways to grow food year-round, especially in areas with harsh winters or limited outdoor space. It offers a unique advantage in situations where going outdoors may be difficult or dangerous, allowing for food production within the safety and comfort of your home.

Key Types of Indoor Gardens:

Herb Gardens: Growing herbs like basil, parsley, cilantro, and thyme on windowsills or in small containers provides fresh seasonings and medicinal herbs. These are some of the easiest and quickest plants to grow indoors and are perfect for adding flavor to meals during an emergency.

- **Microgreens:** Microgreens are young vegetable greens that are harvested just after the first true leaves have developed. These include radish, sunflower, pea shoots, and arugula. They are packed with nutrients and can be grown in small containers in less than two weeks, making them an ideal option for an indoor garden.
- **Leafy Greens:** Lettuce, spinach, kale, and other leafy greens are easy to grow indoors, provided they get enough light. These vegetables are rich in vitamins and are relatively simple to care for in containers or small garden beds.
- **Vertical Gardens:** Utilizing vertical space in your home can significantly increase your indoor gardening potential. Vertical gardening systems use structures like trellises, hanging planters, or wall-mounted shelves to grow plants, maximizing space and efficiency.
- **Hydroponic Systems:** Hydroponics is the method of growing plants in a water-based solution instead of soil. It requires careful monitoring of pH levels and nutrient content but is a highly efficient way to grow food indoors. Hydroponic gardens can produce a variety of crops such as lettuce, spinach, tomatoes, cucumbers, and herbs.
- **Aquaponics:** Combining aquaculture (raising fish) with hydroponics, aquaponics systems rely on the symbiotic relationship between plants and fish. The fish produce waste that provides nutrients for the plants, while the plants help filter the water for the fish. Aquaponics can produce both fish and vegetables, making it a valuable system for food sustainability.

Essential Tools for Indoor Gardening:

- **Grow Lights:** Since many homes lack sufficient natural sunlight, especially in the winter, grow lights are a necessity. Full-spectrum LED grow lights are energy-efficient and can be used to simulate sunlight for plants, encouraging healthy growth.
- **Containers and Pots:** Plants need containers that provide good drainage and sufficient space for root growth. Choosing the right size pot for each plant is crucial for its health.
- **Soil or Growing Medium:** While you can use traditional soil for most indoor plants, hydroponic systems require an alternative growing medium like coconut coir or perlite.
- **Watering System:** An efficient watering system, such as self-watering pots or drip irrigation, helps ensure your plants receive the proper amount of water without becoming over-saturated or dry.

Tips for Successful Indoor Gardening:

- Place plants in areas that receive at least 4-6 hours of indirect sunlight a day, like windowsills, or invest in grow lights.
- Use compact or dwarf varieties of plants that are suited for small indoor spaces.
- Keep indoor gardens clean by regularly removing dead leaves and debris to prevent mold or pests.
- Avoid overcrowding plants; allow space for air circulation to minimize the risk of disease.

3. Outdoor Gardening for Food Security

Outdoor gardening offers more room for growing a larger variety of crops and can yield larger harvests. While it may require more effort and resources, the benefits of outdoor gardening are immense, especially for those with enough space. Outdoor gardening also allows for the use of composting and natural fertilizers, which can contribute to a sustainable cycle of food production.

Key Types of Outdoor Gardens:

- **Raised Bed Gardens:** Raised beds offer an easy and efficient way to grow a wide variety of vegetables, fruits, and herbs. These beds are especially useful for people with poor soil quality or limited space. Raised beds also offer better drainage, are easier to maintain, and help protect plants from pests.
- **Container Gardens:** If you don't have a large yard, you can still grow food in containers, such as pots, barrels, or even recycled materials. Container gardening is an excellent option for people in urban areas with limited space.
- **Greenhouses:** A greenhouse allows you to extend your growing season, offering a controlled environment where plants can thrive year-round. Greenhouses can be small or large and are especially beneficial for growing more delicate crops that need warmer conditions.
- **Traditional Row Gardens:** For those with larger spaces, traditional row gardens allow for the planting of a wide range of crops in long, organized rows. These gardens are great for growing everything from root vegetables to beans, corn, and squash.

Essential Outdoor Gardening Tools:

- **Shovels and Spades:** For digging and turning over soil, a sturdy shovel or spade is essential. This tool is necessary for creating raised beds, planting rows, and preparing your garden.
- **Garden Hose and Irrigation Systems:** Proper irrigation is crucial for outdoor gardens, especially in times of drought. Installing an irrigation system or using soaker hoses can help conserve water while providing your plants with consistent hydration.
- **Composters:** Composting is an excellent way to recycle organic waste into nutrient-rich soil for your garden. Setting up a composting system will help you create fertile soil and reduce the need for chemical fertilizers.
- **Mulch:** Mulching helps retain moisture in the soil, suppresses weeds, and regulates soil temperature. It's especially helpful in hot climates and for maintaining healthy soil in the long term.

Tips for Successful Outdoor Gardening:

Plan your garden layout based on the amount of sunlight and space available. Some plants need full sun, while others thrive in partial shade.

Rotate crops each season to prevent soil depletion and reduce the risk of pests and diseases.

Use companion planting to naturally repel pests and encourage healthy plant growth. For example, planting basil next to tomatoes can improve flavor and repel insects.

Keep an eye on local weather conditions and be prepared to protect your plants from extreme temperatures, frost, or excessive rain.

4. Integrating Indoor and Outdoor Gardening

For a truly sustainable and resilient food supply, integrating both indoor and outdoor gardening methods can be highly effective. For example, you can grow your leafy greens and herbs indoors, while reserving outdoor space for root vegetables, fruiting plants, and larger crops.

Here's how you can integrate both:

Indoor Gardens for Quick Harvests: Use indoor gardening for crops that grow quickly, such as microgreens, herbs, and salad greens. These can supplement your outdoor garden harvests and provide food in the early spring or late fall when outdoor crops are not yet ready.

Outdoor Gardens for Long-Term Sustainability: Outdoor gardens can be used for long-term storage crops like potatoes, onions, carrots, and beans, which can be preserved for use throughout the year.

Transplanting: You can start seeds indoors and then transplant them outdoors once the weather allows. This gives you a head start on your growing season and allows you to grow more varieties than you could otherwise.

Whether indoors or outdoors, gardening for sustenance is an incredibly valuable skill for any prepper or survivalist. Indoor gardening allows you to grow food in small spaces and during winter months, while outdoor gardening provides the space and variety needed for long-term food security. By utilizing a combination of both methods, you can create a sustainable, reliable food source that will help you and your family survive and thrive during any crisis.

Managing Waste and Sanitation in Prolonged Crises

In any extended crisis scenario, managing waste and sanitation is one of the most crucial components of maintaining both health and morale. Whether it's a natural disaster, a civil unrest situation, a pandemic, or a grid-down emergency, the ability to handle waste effectively can prevent the outbreak of diseases, ensure that basic hygiene is maintained, and help create a livable environment for you and your loved ones. Poor sanitation and waste management can result in serious health issues such as cholera, dysentery, and other infectious diseases, making it essential for preppers to plan and implement strategies for waste disposal well in advance.

In this subchapter, we'll discuss practical methods and strategies for managing human waste, food waste, and general garbage, along with techniques for maintaining cleanliness and hygiene in long-term survival situations. We'll cover everything from composting toilets to waste filtration systems and provide you with the knowledge needed to prepare your home or shelter for prolonged emergencies.

1. Understanding the Health Risks of Improper Waste Management

Improper waste management in a crisis can have catastrophic consequences. During disasters, infrastructure such as sewage systems, waste collection services, and water treatment plants often become inoperable. In these situations, human waste and garbage can quickly accumulate and pose significant health risks, including:

- **Contaminated Water Supply:** If waste is not properly handled, it can contaminate nearby water sources, making drinking water unsafe and contributing to the spread of waterborne diseases like cholera and typhoid.
- **Spread of Disease:** Improper waste disposal leads to the accumulation of bacteria, viruses, and pathogens, which can easily spread among people. Stagnant water and waste piles provide breeding grounds for mosquitoes, which can transmit diseases like malaria and dengue fever.
- **Foul Odors and Mental Health:** The presence of waste without proper disposal can create an unbearable stench, lowering morale and causing stress. Mental health issues such as anxiety and depression can increase due to the unhygienic conditions and discomfort.

In the face of these risks, effective waste and sanitation management not only safeguards health but is vital for sustaining long-term survival.

2. Types of Waste in Prolonged Crises

Waste in a survival situation generally falls into three broad categories: human waste, food waste, and general waste. Each type of waste requires a specific approach for safe disposal and sanitation.

Human Waste: Human waste, or excrement, is the most immediate and concerning type of waste during prolonged crises. Inadequate management of human waste can result in dangerous contamination of drinking water and cause the spread of disease.

- **Food Waste:** Leftover food, spoiled food, and packaging materials need to be handled with care to prevent attracting pests, contaminating the environment, or becoming breeding grounds for bacteria and vermin.
- **General Waste:** Includes everything from plastic, metal, and glass packaging to other household waste materials. In a survival scenario, waste buildup can rapidly create environmental hazards, attract scavengers, and degrade living conditions.

3. Managing Human Waste in Prolonged Crises

Handling human waste in emergencies can be tricky, but with the right systems in place, it is manageable. The most critical consideration is maintaining hygiene to prevent the spread of disease.

a. Composting Toilets

One of the most sustainable and practical methods of handling human waste in a prolonged crisis is a composting toilet. These systems break down human waste into compost that can be safely used for gardening after sufficient time has passed. Composting toilets do not require water or sewage systems, making them perfect for off-grid or emergency situations. They typically work by using a combination of aerobic decomposition and materials like sawdust, peat moss, or coconut coir to absorb moisture and neutralize odors.

Benefits:

- **Water Conservation:** Composting toilets use little to no water, which is crucial during water shortages.
- **Sustainability:** The composted waste can be used to fertilize plants, creating a closed-loop system.
- **Odor Control:** With proper maintenance and the right balance of materials, composting toilets are odor-free.

Maintenance Tips:

- Regularly add carbon-rich materials (such as sawdust or straw) to the toilet to keep odors in check and encourage decomposition.
- Empty the compost bin periodically when it is full. It's best to allow compost to fully break down for at least six months before using it as fertilizer.
- Use gloves and disinfect surfaces when emptying the compost to maintain hygiene.

b. Pit Latrines

If a composting toilet is not an option, pit latrines are a common and effective solution. These are simple, often shallow, holes in the ground where waste is deposited and then covered. A key advantage is that they don't require water or plumbing.

Important Considerations:

Location: Choose a location for the pit latrine away from water sources to prevent contamination.

Covering: Always cover human waste with dirt or ash immediately after use to prevent odors and contamination.

Depth: A deep pit (at least 4-6 feet) will help minimize the chance of contamination of the surrounding environment.

Disposal: When the pit becomes full, it must be covered completely and another pit should be dug.

c. Portable Toilets

For those who have limited space or who are confined to smaller shelters, portable toilets can be a viable option. These are easy to set up, operate, and store, and they typically use waste bags or chemicals to help neutralize waste.

Advantages:

- Easy to use and transport.
- Do not require digging or complex installations.
- Some models include waste deodorizing features.

Challenges:

- Requires regular bag replacements or disposal.
- Must be disposed of properly to avoid contamination.

d. Biodegradable Waste Bags

In emergencies, biodegradable waste bags can be used for sanitation purposes. These are particularly useful for situations where access to toilets is limited. They contain enzymes that help break down waste and neutralize odors. These bags can then be buried or disposed of according to safety protocols.

4. Managing Food Waste

Food waste can be another significant challenge during prolonged crises. Without proper disposal, food waste can attract pests, rodents, and insects, and contribute to a sanitary nightmare. Here are some ways to manage food waste effectively:

a. Composting Food Scraps

Composting is an excellent way to recycle food scraps and organic waste into nutrient-rich soil. By creating a compost pile or bin, you can turn leftover food into compost that can be used in your garden to grow new crops. Avoid composting meat, dairy, or oils, as they can attract pests.

Tips for Composting:

- Maintain a balance of "greens" (wet materials like fruit and vegetable scraps) and "browns" (dry materials like leaves, straw, or shredded paper).
- Aerate the compost pile regularly by turning it to provide oxygen and help break down organic matter.
- Keep the compost pile moist but not too wet, and cover it to prevent it from drying out.

b. Secure Storage

In situations where composting is not possible, secure food waste storage is necessary to prevent contamination and the spread of bacteria. Use sealed bags or containers to store food scraps until they can be disposed of safely. Ensure that these containers are kept in cool, dry locations away from your living areas.

5. General Waste Disposal

In a prolonged crisis, managing general waste (non-organic materials like plastic, metal, and paper) is also critical. It's important to have a system for storing and disposing of garbage in a way that doesn't attract pests or compromise hygiene.

a. Garbage Bags and Sealed Containers

Storing non-organic waste in garbage bags or sealed containers helps prevent odor and pest attraction. Be sure to dispose of bags and containers regularly, especially if you're using them to store waste that could decompose or breed bacteria.

b. Recycling and Reuse

In a survival situation, reducing waste by reusing and recycling materials can be highly beneficial. Consider repurposing items like glass jars, metal cans, and plastic bottles for storage or gardening purposes. If there's no recycling infrastructure, create your own makeshift systems for separating and managing waste.

6. Hygiene and Cleaning Solutions

Keeping yourself and your shelter clean during a prolonged crisis is essential for maintaining physical and mental health. In addition to waste management, sanitation practices like washing hands, cleaning surfaces, and disinfecting water are crucial.

a. Handwashing Stations

A simple handwashing station can be set up using water, soap, and a clean container. This should be done regularly, especially after using the toilet, handling food, or cleaning up waste. A foot-pump handwashing station is ideal as it allows you to use water without touching the spigot, maintaining hygiene.

b. Disinfectants and Sanitizers

In the absence of running water, sanitizing wipes, alcohol-based hand sanitizers, or homemade disinfectants (made from bleach or vinegar) can be used to clean surfaces. Regular cleaning can help prevent the buildup of germs and bacteria that can lead to infections or illnesses.

c. Regular Cleaning of Living Spaces

Maintaining cleanliness in your shelter will go a long way in boosting morale and promoting health. Regularly clean floors, surfaces, and sleeping areas to minimize the risk of infection and pest infestations.

Managing waste and sanitation during prolonged crises requires preparation, knowledge, and the right tools. Whether through composting toilets, waste bags, or proper storage techniques, ensuring that waste is handled in a safe and hygienic manner will be critical to maintaining both your health and the health of those around you. In a survival scenario, these simple yet effective practices will help reduce the spread of diseases, prevent environmental contamination, and promote a safer, healthier living environment.

9. DIY SURVIVAL PROJECTS

Building a Rocket Stove

A rocket stove is an efficient and sustainable way to cook food and heat your home, especially in emergency or off-grid situations. It is designed to burn small amounts of fuel (such as twigs, branches, or wood pellets) at extremely high efficiency, producing minimal smoke and maximizing heat output. The rocket stove's unique design, which channels heat and air in a manner that creates a clean, intense burn, makes it a popular choice among preppers, off-gridders, and survivalists who need to conserve fuel and minimize their environmental footprint.

In this section, we'll cover the basic principles behind rocket stoves, the materials you'll need, and step-by-step instructions on how to build a rocket stove at home. We'll also explore the advantages of rocket stoves, how to use them effectively, and how to maintain them for long-term use.

1. The Science Behind the Rocket Stove

The rocket stove operates on a principle called the "rocket combustion" process, which involves a vertical combustion chamber (or "rocket tube") and an insulated, horizontal combustion chamber that directs the flow of heat and gases in a highly efficient manner. The key elements of the design include:

- **Insulation:** The rocket stove features high-temperature insulation (often made from firebricks, clay, or other heat-retentive materials) that keeps the combustion area hot and ensures that the stove burns efficiently. This prevents heat loss, allowing the stove to run at a high temperature with a small amount of fuel.
- **Efficient Airflow:** The stove is designed to funnel oxygen into the fire through a narrow channel, creating a hotter and more complete combustion process. This means less fuel is needed to generate the same amount of heat, and there's less smoke.
- **Vertical Combustion Tube:** The fuel burns in a vertical column, where hot gases rise rapidly. This helps the stove burn at a higher temperature and creates a more efficient heat exchange.

2. Materials Needed to Build a Rocket Stove

Before you begin building your rocket stove, it's important to gather the necessary materials. Most rocket stoves can be made from common materials that are easy to find in most parts of the world. Here's what you'll need:

a. Fuel

- Small twigs, sticks, or branches
- Wood pellets or small pieces of firewood
- Other biomass materials (dry leaves, straw, or corn stalks)

b. Building Materials

- **Metal Cans:** Large metal cans, such as empty paint cans or large tuna fish cans, are often used for the body of the stove. These can be easily found and repurposed.
- **Firebricks or Clay:** Firebricks are heat-resistant and help retain the stove's heat. Alternatively, you can use clay or a mix of sand and ash to make your own insulating material.
- **Steel Tubing or Pipe:** This is used for the vertical combustion chamber and the air intake system. Steel pipe helps create a clean burn by directing the airflow.
- **Concrete or Mortar:** A heat-resistant mortar or concrete mix will be necessary to secure and seal parts of the stove.

c. Tools

- **Metal-cutting tools:** A saw or grinder to cut metal cans or pipes
- **Drill:** To create air vents or holes in cans or pipes for air flow
- **Gloves and safety glasses:** To protect yourself while handling metal or cutting tools

3. Step-by-Step Instructions for Building a Basic Rocket Stove

Building a rocket stove doesn't require advanced skills, and the materials are often easy to find. Here's a simple step-by-step guide for building a basic rocket stove using metal cans and a steel pipe.

Step 1: Create the Base

Start by selecting a sturdy surface for your stove. This could be a heat-resistant tile, concrete block, or another fireproof base.

Take a large metal can (a paint can or large coffee can works well) and cut a hole in the side near the bottom, about 3 inches wide. This will serve as the fuel intake. The hole should be placed at a slight upward angle so that the twigs or sticks can easily be fed into the stove.

Cut a small hole near the top of the can on the opposite side to serve as the exhaust for the smoke. This hole should be about 1 inch in diameter.

Step 2: Create the Vertical Combustion Chamber

Use a steel pipe or a smaller metal can for the combustion chamber. The combustion chamber should be slightly taller than the base can.

Cut a hole in the bottom of the vertical pipe to align with the fuel intake hole of the base can. This allows the fuel to feed into the combustion chamber.

If you are using a larger can for the combustion chamber, you can attach a steel pipe or another metal tube to the top, forming a chimney to direct the smoke and heat.

Step 3: Build the Insulating Jacket

Create a heat jacket to insulate the combustion chamber and prevent heat loss. This can be done by surrounding the combustion chamber with firebricks or clay bricks, creating an insulating layer.

If using firebricks, stack them around the stove to create a small enclosure around the combustion chamber. Leave enough space around the combustion tube for airflow.

Once the combustion chamber is insulated, you can build a simple stovetop using another metal sheet or flat steel plate to cover the top. This will act as the cooking surface.

Step 4: Add Air Vents

To ensure efficient burning, the rocket stove needs a continuous airflow to feed the fire with oxygen. Drill several small holes (air vents) around the stove at various levels, especially around the intake area. These holes should allow air to flow into the combustion chamber, ensuring that the fuel burns hot and clean.

Step 5: Final Assembly

Once all components are cut and in place, you can begin assembling the stove. Secure the parts with mortar or heat-resistant glue. Make sure the joints are airtight to maintain a hot, efficient burn.

Attach the vertical combustion tube to the base and make sure the fuel intake hole is aligned properly.

Use metal fasteners or mortar to fix the combustion chamber into place.

Add the stovetop cover and make sure it's stable and secure.

4. Benefits of Building a Rocket Stove

There are several advantages to using and building a rocket stove, particularly in off-grid situations or emergencies:

a. Fuel Efficiency

Rocket stoves are extremely fuel-efficient. Because they burn at such high temperatures, they use much less fuel compared to traditional wood stoves or campfires. Small twigs and sticks are often enough to generate significant heat, making rocket stoves an excellent option for places with limited access to large amounts of firewood.

b. Low Smoke Production

The combustion process in a rocket stove is highly efficient, which means that there is much less smoke produced compared to traditional stoves. This makes them ideal for use in areas where air quality is a concern, or where smoke signals could attract unwanted attention in a survival scenario.

c. Sustainability

Since rocket stoves use small, renewable fuels like twigs, leaves, and branches, they are an environmentally friendly option for cooking and heating. You don't need to rely on gas or electricity, making them an ideal solution for off-grid living.

d. Portability

Once built, rocket stoves can be easily transported. Many designs are compact enough to be taken on camping trips or used as emergency backup cooking devices. Some portable models can be easily disassembled and packed for travel.

5. Maintaining and Using Your Rocket Stove

To ensure your rocket stove lasts and remains efficient, regular maintenance is essential:

a. Keep it Clean

After each use, clean out any leftover ash or debris to ensure optimal airflow and prevent clogs.

b. Regularly Check for Wear and Tear

Check the combustion chamber for any signs of rust or damage. The high heat generated by the stove can cause metal parts to deteriorate over time, so it's important to inspect the stove regularly.

c. Proper Storage

If using your rocket stove in an off-grid or emergency setting, store it in a dry area to prevent rust. Ensure that the stove is stored away from flammable materials when not in use.

Building a rocket stove is an excellent way to ensure that you have a reliable, efficient, and sustainable cooking and heating source during emergencies or off-grid living situations. By using small, renewable fuel sources and minimizing smoke, a rocket stove provides a practical solution to the challenge of cooking without access to electricity or gas. With some basic materials and a little time, you can build your own rocket stove and gain the confidence and skills to thrive during crises.

Constructing Rainwater Collection Systems

Rainwater harvesting is a time-tested and sustainable solution for gathering and utilizing water in areas where access to potable water is limited or unreliable. For preppers and those living off-grid or in emergency situations, setting up a rainwater collection system can provide a valuable resource for drinking, cooking, cleaning, irrigation, and even flushing toilets. This section will guide you through the process of constructing your own rainwater collection system, covering the essential components, the design, and how to properly maintain the system to ensure a reliable water supply.

1. Why Rainwater Collection Is Essential for Preppers

In a survival situation or during long-term emergencies, water is a critical resource. As climate change increases the frequency and intensity of extreme weather events, such as floods, droughts, and hurricanes, many people are turning to rainwater collection systems to secure an independent and sustainable water source.

Benefits of rainwater harvesting include:

Water Independence: Rainwater collection allows you to become less dependent on municipal water systems or well water, which may be unreliable in the event of natural disasters or infrastructure breakdowns.

Cost-Effective: Collecting rainwater can lower your water bills, especially in regions where water usage is metered.

Environmentally Friendly: It reduces runoff, prevents erosion, and helps replenish local groundwater supplies. In urban areas, rainwater harvesting can also help prevent flooding and reduce the strain on stormwater systems.

Versatility: Collected rainwater can be used for a variety of purposes, from drinking and cooking to irrigation and sanitation, making it a versatile addition to your preparedness plan.

2. Key Components of a Rainwater Collection System

To build an efficient rainwater collection system, you need to understand the basic components that make up the system. Each part plays a vital role in capturing, filtering, storing, and distributing the water you collect.

a. Catchment Surface

The catchment surface is where rainwater first falls and is collected. In most residential rainwater systems, the catchment surface is the roof. However, roofs made from certain materials, such as tar, asphalt, or painted metals, may not be ideal for water collection due to the potential for contamination.

Material Considerations: Ideally, use a roof made of materials such as clay tiles, metal, or corrugated iron, which are easier to clean and don't shed harmful chemicals into the water.

Slope and Size: The roof should have enough slope to ensure rainwater flows efficiently into the gutters. Larger roofs collect more water, but they also require larger storage systems and more maintenance.

b. Gutters and Downspouts

Once the rainwater hits the roof, it needs to be channeled into a storage system. Gutters are installed along the edges of the roof to collect the rainwater and direct it into downspouts. These are vertical pipes that channel the water from the gutters into your storage containers.

Materials: Gutters can be made from galvanized steel, PVC, or aluminum, and they should be strong enough to handle the volume of water during heavy rain.

Cleaning: Regular cleaning of gutters is necessary to ensure they do not clog with leaves or debris, which could block water flow.

Gutter Guards: To reduce the frequency of cleaning, consider installing gutter guards or mesh filters to keep out leaves, twigs, and other debris.

c. First Flush Diverter

The first flush of rainwater after a dry spell can carry contaminants from the roof, such as dirt, bird droppings, and other debris. To prevent this contaminated water from entering your storage tank, you should install a first flush diverter.

How It Works: The diverter captures the first few gallons of rainwater, which is then directed away from the storage system. After this initial flush, the cleaner rainwater is directed into the storage tank.

Maintenance: Regularly clean the first flush system to ensure that it continues to work effectively and doesn't become clogged with debris.

d. Storage Tanks

The water collected from the roof and downspouts is directed into a storage tank or cistern. The size of the tank depends on the average rainfall in your area and the amount of water you expect to use.

Tank Materials: Storage tanks can be made from a variety of materials, including plastic, fiberglass, concrete, and metal. Plastic tanks are common due to their durability, low cost, and ease of installation.

Capacity: The tank's capacity is measured in gallons or liters. A typical household may require anywhere from 1,000 to 5,000 gallons of storage, depending on the number of people and their water usage.

Placement: The storage tank should be placed on a level surface, preferably elevated, to allow gravity to help distribute the water into the system. You can build a small platform using bricks or concrete blocks to elevate the tank.

e. Filtration System

For drinking water, it's essential to filter and purify the rainwater before use. A good filtration system removes contaminants like dirt, debris, and pathogens, ensuring that the water is safe to drink.

First Stage: A basic pre-filter or mesh screen can be installed at the entry point of the storage tank to capture larger debris.

Secondary Filtration: For further filtration, use activated carbon filters, sand filters, or ceramic filters. These remove smaller particles and chemicals that may be present in the water.

Advanced Filtration: For higher-level purification, especially in areas with potential for pathogens, a UV filter or reverse osmosis system can be added to kill harmful microorganisms.

f. Water Distribution

Once the rainwater is stored and filtered, it needs to be distributed for use. This can be achieved through a pump system that moves the water from the storage tank to your home or garden.

Manual Pumps: In off-grid settings, hand-operated pumps can be used to manually transfer water.

Electric Pumps: For more convenient systems, electric pumps can be used, though you should have a backup power source in case of a power outage.

3. Designing Your Rainwater Collection System

When designing a rainwater harvesting system, several factors must be considered to ensure maximum efficiency:

a. Assess Water Needs

Daily Consumption: Consider how much water your household uses on a daily basis. A family of four might need anywhere from 80-100 gallons of water per day, depending on local habits.

Rainfall Patterns: Check local climate data to understand how much rainfall your area typically receives. This will help you size your collection system appropriately.

Backup Water Sources: In case of prolonged dry periods, consider how your rainwater collection system can be supplemented with other water sources, like well water or a nearby stream.

b. Plan for Overflow

In areas with heavy rainfall, your collection system will need to accommodate overflow to prevent damage. Overflow outlets on your tanks should be directed away from the foundation of your home, and a properly designed system will redirect excess water safely.

c. Size the System

Gutter Size: Larger gutters can catch more rain, so consider upgrading your gutters to handle higher volumes of water if you live in an area that experiences heavy rainfall.

Tank Size: Calculate the capacity of your storage tank based on your daily water usage and the rainfall in your area. A larger tank means fewer trips to refill, but it also requires more space.

4. Maintenance and Troubleshooting

Maintaining your rainwater collection system ensures its longevity and functionality:

Clean Gutters and Filters: Inspect and clean your gutters, first flush diverters, and filters regularly, especially after storms or periods of heavy rain.

Check for Leaks: Look for leaks or cracks in your storage tank or pipes. Small leaks can lead to significant water loss over time.

Inspect Pumps: Ensure that your water pump is in good working order, and regularly check the connections to avoid dry runs.

Constructing a rainwater collection system is a valuable skill for preppers, off-grid dwellers, and those living in areas with unreliable water sources. With a little investment in time and resources, you can create a system that provides a continuous and sustainable water supply for drinking, cooking, and everyday needs. By considering factors like system size, filtration, and maintenance, you can ensure that your rainwater collection system functions efficiently, offering you a reliable alternative water source in emergencies or long-term survival scenarios.

Creating Homemade Air Filtration Units

In situations where air quality is compromised—whether due to natural disasters, environmental pollution, or chemical threats—having a reliable air filtration system can be a matter of survival. Commercial air purifiers can be expensive and may not always be accessible in an emergency, but creating your own homemade air filtration units can provide an effective, low-cost solution. This subchapter will guide you through the process of building your own air filtration units using readily available materials, ensuring that you and your family can breathe safely in any emergency.

1. Why You Need an Air Filtration System

Airborne pollutants—such as dust, smoke, allergens, volatile organic compounds (VOCs), and even toxic gases—pose serious health risks, especially in confined spaces. During emergencies like wildfires, industrial accidents, or in a post-nuclear event scenario, air quality can quickly deteriorate. Without proper filtration, inhaling these contaminants can lead to respiratory issues, allergies, headaches, fatigue, and long-term health problems.

Health Risks: Polluted air can trigger asthma, worsen lung conditions like COPD (Chronic Obstructive Pulmonary Disease), and lead to irritation of the eyes, nose, and throat.

Survival Importance: Clean air is essential for overall survival in an emergency. In situations such as wildfires or chemical spills, filtering out harmful particles can be a life-saving measure.

2. Basic Components of an Air Filtration System

To build an effective homemade air filtration system, it's important to understand the key components that make up a functional unit. These systems typically rely on layers of filtering media, each designed to capture different types of airborne pollutants.

a. Fan (Airflow Source)

A fan is the heart of any air filtration system. It pulls in contaminated air and forces it through a series of filters before releasing purified air back into the room.

Type of Fan: The fan should be able to create enough airflow to circulate air through the filters effectively. Box fans or computer fans can be used, but larger, higher-quality fans (often found in air purifiers) are more efficient at moving air.

Fan Size: Ensure the fan is appropriately sized for the room where it will be used. Larger spaces will require bigger or multiple fans to achieve the desired airflow.

b. Filter Media

The filter media is the primary layer of the filtration system and determines how effective the system is at trapping contaminants. There are several types of filters, each designed to target specific types of pollutants.

HEPA Filters: High-Efficiency Particulate Air (HEPA) filters are highly effective at trapping particles like dust, pollen, smoke, and other microscopic airborne contaminants. HEPA filters capture particles as small as 0.3 microns with a 99.97% efficiency rate.

Activated Carbon Filters: Activated carbon is effective at removing volatile organic compounds (VOCs), odors, and gases. This filter works by adsorbing contaminants onto the surface of the activated carbon, making it ideal for neutralizing odors from smoke or chemicals.

Pre-Filters: A pre-filter is a less dense filter used to trap larger particles like dust and hair. Using a pre-filter can help extend the life of the more expensive HEPA filter by preventing it from getting clogged too quickly.

c. Housing (Filter Housing or Box)

The housing holds the filters and fan in place and directs the airflow through the system. It can be as simple as a wooden or plastic box, depending on the materials you have at hand.

Material Considerations: Use durable materials like plywood, thick cardboard, or plastic for the housing. The key is to ensure that the fan and filter are tightly sealed into the box to prevent unfiltered air from bypassing the system.

Sealing: Proper sealing around the edges of the filters and housing is essential to ensure that air can only pass through the filters, preventing contaminants from escaping.

3. Step-by-Step Guide to Building a Homemade Air Filtration Unit

Now that you understand the basic components, let's walk through the steps of building a homemade air filtration unit.

Materials Needed:

- Box fan (12" to 20" diameter, depending on room size)
- HEPA filter (size compatible with your fan)
- Activated carbon filter (optional, for chemical and odor removal)
- Duct tape or strong adhesive
- Cardboard or plywood for housing
- Scissors or utility knife
- Plastic or metal fasteners (optional, for durability)

Step 1: Preparing the Fan

Start by selecting a box fan or another type of fan that you can attach a filter to. Ideally, the fan should have an intake that can be sealed around the edges to prevent unfiltered air from passing through.

Remove the fan cover if necessary, and prepare the front intake area for mounting the filter. You may need to cut the cardboard or plywood to create a custom housing for the fan.

Seal the edges of the fan using duct tape to ensure there are no air gaps. This is essential for efficient filtration.

Step 2: Adding the HEPA Filter

Cut the HEPA filter: If the filter is too large for the fan, trim it to the correct size. The goal is for the filter to fit snugly on the fan intake without gaps.

Attach the filter: Use duct tape or adhesive to secure the filter onto the fan intake. Make sure the filter is firmly in place and that there is no air leakage around the edges.

Step 3: Optional Addition of Activated Carbon Filter

If you want to remove odors, VOCs, or gases from the air, consider adding an activated carbon filter. These can be placed directly before the HEPA filter or behind it in the airflow path.

Placement: The activated carbon filter should be placed so that air flows through it before reaching the HEPA filter. This allows the carbon to absorb gases and odors before the air is filtered for particles.

Step 4: Build the Housing

Construct the frame: If necessary, build a frame to hold the fan and filter in place. This could be a simple wooden box or a cut-out from cardboard. The frame should be large enough to securely house the fan while allowing for the air to flow freely through the filter.

Seal the housing: Ensure all edges are tightly sealed with duct tape to prevent any air from bypassing the filters.

Step 5: Testing and Adjustment

Once the fan and filters are securely in place, plug in the fan and test the system. Check for any air leaks around the filter edges, and make adjustments to improve the seal if needed.

Airflow: Ensure that the fan is pushing air through the filters and out of the unit. You may need to adjust the fan speed or placement to optimize airflow.

4. Maintenance and Troubleshooting

Once your homemade air filtration system is up and running, regular maintenance is essential to ensure it continues to function effectively.

a. Filter Replacement

HEPA Filter: Depending on the level of air pollution, you will need to replace the HEPA filter every 6 months to a year. If you're using the filter in a particularly smoky or polluted area, you may need to replace it more often.

Activated Carbon: Activated carbon filters should be replaced more frequently, as they lose their effectiveness once they are saturated with contaminants.

b. Fan Maintenance

Clean the fan blades and motor periodically to ensure that the fan is running smoothly. Dust and debris can accumulate on the fan over time, reducing its efficiency.

c. System Performance

Regularly check the system for any signs of malfunction or reduced air quality. If you notice that the air isn't as clean as it should be, inspect the filters, fan, and seals to make sure everything is working properly.

Creating your own homemade air filtration system is a cost-effective way to ensure that you and your family have access to clean air in an emergency. With a few simple components—like a fan, HEPA filters, and activated carbon—you can build a filtration unit that helps protect you from harmful airborne contaminants. By regularly maintaining the system and replacing filters as needed, you can rely on your homemade air filtration unit to provide safe, breathable air in any crisis situation.

Crafting Emergency Lighting Solutions

When disaster strikes and the power grid fails, one of the immediate needs you'll face is the lack of light. Whether it's a storm, earthquake, or even a grid-down situation caused by civil unrest, having a reliable and sustainable source of light becomes essential. In the absence of electricity, relying on candles or flashlights alone can be insufficient and unsafe. Therefore, crafting emergency lighting solutions is not only a necessity but also an opportunity to improve your home preparedness plan and create long-lasting, versatile systems that can ensure safety and comfort in the dark.

This subchapter will cover several DIY solutions, tips, and strategies for crafting emergency lighting systems that are practical, cost-effective, and easy to implement, offering you the ability to create light when you need it most.

1. Understanding the Importance of Emergency Lighting

Why It's Vital

In an emergency, lighting is not just about convenience; it's about survival. Poor visibility can lead to accidents, disorientation, and even injuries. Whether you're navigating a power outage, staying put during a storm, or trying to keep watch during an emergency, a lack of light can severely hinder your ability to function and make sound decisions.

Safety: Without adequate lighting, you increase your risk of tripping, falling, or accidentally injuring yourself. It can also make it more difficult to monitor external threats.

Security: Lighting acts as a deterrent against potential intruders, especially during extended power outages when the darkness can be used by opportunistic criminals.

Comfort and Morale: Light helps maintain a sense of normalcy, reducing anxiety and boosting mental and emotional well-being. It can also help maintain a routine during extended blackouts or disruptions.

2. Types of Emergency Lighting Solutions

There are various types of emergency lighting solutions you can craft to suit your needs. Depending on your resources and preferences, you can create lighting that is powered by a variety of methods—such as solar, battery, and even hand-crank systems. Here are some practical options:

a. Solar-Powered Lights

Solar-powered lighting solutions are an excellent choice for emergencies, as they provide a renewable and independent source of light.

Solar Lanterns and Flashlights: These are small, portable units that can charge during the day and provide hours of light at night. Many modern solar-powered lanterns also feature built-in USB ports, allowing you to charge other devices in emergencies.

DIY Solar-Powered Lights: You can create a solar-powered lighting system by using photovoltaic (PV) cells, rechargeable batteries, and LED lights. Place the PV cells in a location with direct sunlight to charge the batteries during the day.

Advantages: Solar lights are renewable and require minimal maintenance. They are perfect for long-term power outages as they work independently of the electrical grid.

b. Hand-Crank Powered Lights

Hand-crank lights are another fantastic option for emergencies where battery power may be scarce, and sunlight may not be sufficient.

How They Work: Hand-crank lights are powered by manual energy. A crank mechanism generates electricity that is stored in the unit's internal battery, which powers an LED bulb. Some units provide both lighting and the ability to charge small devices via USB.

Best for Short-Term Needs: Hand-crank lights are not designed to provide hours of illumination continuously but are great for short-term lighting in emergencies. They're reliable because they don't rely on batteries that can lose charge over time.

How to Craft One: Building a simple hand-crank generator can be done with a small DC motor, a flywheel, and a battery storage unit. These components can be assembled into a system that stores energy for later use, powering an LED light when cranked.

c. Battery-Powered LED Lights

Battery-powered LED lights are among the most versatile, affordable, and accessible options for emergency lighting.

Flashlights: Traditional flashlights can be a lifesaver in emergencies, but modern LED flashlights are far more energy-efficient and longer-lasting than older incandescent ones. They use far less power to produce brighter light, which means you can rely on them for extended periods.

Battery-Powered Lamps: If you prefer an ambient light source rather than a flashlight, you can create a DIY lantern using LED strips or bulbs powered by common household batteries. Many battery-powered LED lamps also come with dimmer controls, allowing you to adjust the light intensity as needed.

Building Your Own: You can craft a simple LED lantern by attaching LED strips to a plastic container or mason jar, with a battery holder attached to the bottom. A small switch will allow you to control the light.

d. Candles and Oil Lamps

While not as modern or efficient as other options, candles and oil lamps have been long relied upon for emergency lighting, and they still have a place in emergency preparedness plans.

Candles: These are easy to store and cheap to purchase. While candles don't require electricity or batteries, they do have fire risks and a limited burn time. Store them in a safe, ventilated location, and consider flameless candles as a safer alternative.

DIY Oil Lamps: Oil lamps are a great option for longer-lasting lighting during an emergency. A simple DIY oil lamp can be made using a small jar (like a mason jar), a wick (made from cotton or other flammable fabric), and a flammable liquid such as lamp oil or vegetable oil. These homemade lamps can provide light for hours.

Safety: Be sure to store candles and oil lamps in safe, stable locations to prevent fires, especially in windy or unstable conditions.

3. How to Build a Simple Solar-Powered Light

Let's break down how you can craft a simple solar-powered light from readily available materials. This project will give you the tools to light up your home in the event of a power failure while utilizing renewable energy.

Materials You'll Need:

- Small solar panel (preferably 5-10W)
- Rechargeable 12V battery
- LED bulb (12V or 5V, depending on the battery)
- Charge controller (optional, for better efficiency)
- Diode (to prevent backflow of electricity)
- Wires and connectors
- Soldering iron and wire (optional)
- Plastic housing (to protect the system from the elements)

Steps:

Connect the Solar Panel to the Battery: The solar panel will be used to charge the battery. Use the diode to prevent reverse current, ensuring that the battery doesn't discharge when it's dark.

Connect the Battery to the LED: Attach the battery to the LED light, making sure you have the appropriate connectors or wires to handle the power. You may need a charge controller to prevent overcharging or over-discharging the battery.

Enclose in a Housing: Use plastic or any waterproof container to house the solar panel, battery, and wiring. This will protect your system from the elements.

Install in a Sunny Location: Place the solar panel outside in an area with direct sunlight to charge the battery during the day. In the evening, your LED light should turn on automatically as the stored energy powers the light.

4. Maintaining Your Emergency Lighting System

Once you've built or gathered your emergency lighting systems, it's important to maintain them so that they'll function when needed the most.

a. Check Battery Life Regularly

Battery-powered systems require regular monitoring to ensure that batteries are still functioning well. Test the batteries periodically to avoid any surprises during an emergency. Store extra batteries in a cool, dry place to maximize their lifespan.

b. Clean Your Solar Panels

Dust and debris can block sunlight from reaching your solar panels. Clean them every few months to maintain optimal charging efficiency.

c. Safety Precautions

Store all lighting systems, particularly candles and oil lamps, in safe areas away from flammable materials. Always keep matches, lighters, and oil in a secure, fireproof container. If you have children in the home, ensure that lighting systems with open flames are kept out of their reach.

Crafting emergency lighting solutions doesn't have to be expensive or complicated. Whether you're relying on solar power, hand-crank systems, or DIY oil lamps, there are numerous ways to ensure that you're never left in the dark during an emergency. By combining these simple solutions with regular maintenance and safety practices, you can keep your home illuminated and secure when the lights go out, giving you the freedom and peace of mind to focus on what matters most: survival and well-being.

10. EMERGENCY MEDICAL PREPAREDNESS

Assembling a Complete First Aid Kit

A well-equipped first aid kit is an indispensable part of any emergency preparedness plan. In a survival situation, whether it's due to natural disasters, accidents, or long-term disruptions, quick access to medical supplies can make the difference between life and death. A complete first aid kit ensures that you can manage minor injuries and medical issues until professional medical help arrives or until you can safely evacuate.

This subchapter will guide you through the process of assembling a comprehensive first aid kit. We'll discuss the essential items that should be included, how to organize them for easy access, and why it's important to tailor your kit to specific needs.

1. The Basics of a First Aid Kit

A first aid kit serves as the foundation for addressing medical emergencies. It's important to focus on a variety of items that cover the most common injuries and conditions you may encounter. These include cuts, burns, sprains, fractures, infections, and other acute medical conditions. Ideally, your first aid kit should be versatile enough to handle minor emergencies as well as more serious medical issues in the absence of professional help.

The American Red Cross and the World Health Organization (WHO) both provide guidelines for creating a well-rounded first aid kit that can cater to different types of emergencies. Below is a comprehensive list of essential items to consider when assembling your first aid kit:

2. Essential Items for a Complete First Aid Kit

a. Bandages and Dressings

Adhesive Bandages (Band-Aids): Various sizes to cover small cuts, abrasions, and blisters. Consider sterile versions for clean and safe protection.

Sterile Gauze Pads and Rolls: For larger wounds that require absorption and protection from dirt and bacteria.

Adhesive Tape: To secure dressings and bandages in place. Medical-grade tape, like paper tape or cloth tape, is often the best for skin safety.

Elastic Bandage (ACE Bandage): For sprains, strains, and to support joints or muscles.

Non-Adherent Pads: For large wounds or burns, these pads prevent sticking to the injury while absorbing any exudates.

Burn Dressing: Gel-soaked burn dressings are vital for cooling and treating burn injuries in the early stages.

b. Antiseptics and Medications

- **Antiseptic Wipes and Alcohol Pads:** To clean and disinfect wounds, cuts, and abrasions before dressing them. These help prevent infection by removing dirt and bacteria.
- **Antibiotic Ointment (e.g., Neosporin):** Apply to minor cuts and abrasions to prevent infection.
- **Hydrocortisone Cream:** For treating rashes, insect bites, and allergic reactions.
- **Aspirin and Acetaminophen:** Pain relievers are essential for managing headaches, muscle pain, or general discomfort during a crisis.
- **Ibuprofen:** This non-steroidal anti-inflammatory drug (NSAID) helps with inflammation, pain, and fever.
- **Antihistamines (e.g., Benadryl):** Useful for allergic reactions to bites, stings, or pollen.
- **Antacid Tablets:** For digestive issues or acid reflux, which might become an issue if food and water supplies are limited.
- **Tweezers:** Essential for removing splinters, ticks, or small foreign objects from the skin.

c. Wound Care and Injury Management

Splints: For immobilizing broken or fractured bones. A few ready-made splints, or improvise with sturdy materials like wooden sticks or rolled-up newspapers.

Instant Cold Packs: For reducing swelling and numbing pain from sprains, burns, or bruises.

Elastic Bandage (e.g., ACE bandage): Can be used to wrap sprained or strained limbs, helping to reduce swelling and provide support.

Finger Splints and Toe Pads: Specifically for immobilizing injured fingers or toes.

d. Medical Tools

Scissors: To cut gauze, tape, or clothing to access wounds.

Thermometer: An important tool to monitor for fever, especially when dealing with infections or viruses.

Gloves: Non-latex gloves should be included for sanitation when treating injuries. These prevent contamination of wounds and minimize your exposure to blood and bodily fluids.

CPR Shield or Face Shield: Essential in case CPR is needed, providing protection for both the responder and the patient.

Eye Wash or Eye Drops: In case of eye injuries or irritations, having a saline solution on hand will allow you to rinse out the eyes safely.

e. Specialized Medical Supplies

Burn Gel and Dressings: Burn injuries, whether from fire, hot surfaces, or chemicals, require immediate attention. Burn gels help cool the burn, while sterile burn dressings prevent infection.

Snake Bite Kit: If you live in an area prone to snakes, include a kit with suction tools and instructions for immediate care.

Anti-Diarrheal Medicine: In a survival situation, dehydration from diarrhea can quickly become fatal. Include over-the-counter anti-diarrheal medication, especially if you're in an area with uncertain water quality.

Aspirin: In the case of heart attacks or chest pain, aspirin can help thin the blood and prevent further clotting until medical attention arrives.

f. Personal Health Items

Prescription Medications: If you or a family member take prescription medications, be sure to include an emergency supply of each. Stock at least a two-week supply in your kit, and check expiration dates regularly.

Inhalers: For anyone with asthma or respiratory issues, include spare inhalers and/or nebulizer medications.

EpiPens: For individuals with severe allergies, especially to food, insects, or medication, an EpiPen is essential in an emergency.

3. Organizing Your First Aid Kit

It's not enough to just throw these items into a box and call it good. A well-organized first aid kit allows for quick access during an emergency when time is of the essence. Here are some tips for effectively organizing your first aid kit:

Use Clear, Labelled Containers: Use clear, resealable plastic bags, small plastic containers, or a compact first aid box. Label sections clearly so that you can easily find what you need in a hurry.

Group Items by Category: Separate items by category (e.g., wound care, medications, tools) and store them together. This will allow you to locate and use items more efficiently.

Add a First Aid Manual: Include a basic first aid manual or instruction card with your kit, which can help you take the correct actions in the event of an emergency.

Keep It Accessible: Store your kit in a place that is easy to access, especially during emergencies. Consider multiple locations, including in your car, garage, and home.

Check Expiration Dates: Periodically check the expiration dates on medications, bandages, and other supplies. Rotate items to keep your kit fresh.

4. Customizing Your First Aid Kit

Different environments and situations may call for specific additions to your first aid kit. Here are a few suggestions based on unique needs:

For Families with Children: Add baby-safe pain relievers, baby bandages, and teething gels.

For Pet Owners: Include pet-safe antiseptics, bandages, and medication for your animals.

For Remote Areas or Off-Grid Survival: Consider adding items such as a field surgical kit, anti-parasitic medications, and more advanced trauma care tools.

5. Regular Maintenance and Updates

To ensure your first aid kit remains ready for use, you must maintain it by checking supplies and replenishing anything that's been used. Replace any expired medications, and test items like bandages, gauze, and thermometers to ensure they're in working order. Regularly updating your kit will keep it functioning when emergencies arise.

Assembling a complete first aid kit is one of the most critical steps in preparing for an emergency. With the right supplies, you'll be equipped to handle a wide range of medical situations, from minor cuts and bruises to more severe injuries that require immediate care. Whether you're staying in place or bugging out, having access to medical supplies and knowing how to use them will ensure that you and your family stay safe during crises. By organizing your kit, customizing it to your unique needs, and maintaining it regularly, you can confidently face whatever comes your way.

Learning Basic and Advanced Medical Skills

In a survival or emergency situation, knowing how to address medical needs effectively is crucial. Whether you are "bugging in" during a natural disaster, preparing for a grid-down scenario, or dealing with the aftermath of civil unrest, having basic and advanced medical skills can mean the difference between life and death.

This subchapter will provide insight into the importance of medical knowledge, the types of skills you should prioritize, and how to acquire both basic and advanced medical training. It will also highlight how these skills can be applied in a variety of survival situations, empowering you to take care of your family and community in times of crisis.

1. The Importance of Medical Skills in Survival Situations

In survival scenarios, access to modern healthcare facilities is often limited, or even nonexistent. Hospitals may be overwhelmed with patients, medical professionals could be unavailable due to disasters, or travel to a medical facility might be impossible because of blockages or security concerns. Therefore, your ability to manage medical situations independently becomes vital.

Being prepared with essential medical skills will also enable you to avoid relying on others, which could be dangerous or impractical. The ability to administer first aid, handle common injuries, recognize symptoms of disease, and even stabilize a life-threatening condition can keep you and your loved ones alive until you are able to receive professional care, if necessary.

2. Basic Medical Skills Everyone Should Learn

Basic medical skills form the foundation for addressing everyday health issues and minor injuries. These skills are essential for self-sufficiency, especially when traditional medical assistance is unavailable. Here's a look at the fundamental medical skills you should focus on learning:

a. Basic First Aid

First aid is the cornerstone of basic medical care and involves providing immediate treatment for common injuries or health conditions. Key skills include:

Wound Cleaning and Bandaging: Knowing how to clean and dress a wound properly prevents infection and promotes faster healing.

CPR (Cardiopulmonary Resuscitation): Learning CPR can save lives in cases of cardiac arrest. Knowing how to properly perform chest compressions and rescue breathing is essential.

Splinting and Fracture Care: Understanding how to immobilize broken bones with improvised materials can prevent further injury and pain.

Burn Care: Treating burns requires immediate action to cool the affected area and prevent infection. Knowing how to apply the right type of dressing for different burn types is vital.

Recognizing Signs of Shock: Recognizing the symptoms of shock (such as rapid breathing, weak pulse, confusion, and cold skin) and taking appropriate steps to stabilize the patient is crucial.

b. Basic Medical Diagnostics

In survival situations, you may need to diagnose common ailments or injuries with limited resources. Developing basic diagnostic skills will help you decide when to treat an injury and when it's necessary to seek more advanced medical attention. Some basic diagnostic skills include:

Taking Vital Signs: Learning to take and interpret vital signs, such as pulse, blood pressure, body temperature, and respiration rate, will give you a better understanding of someone's health status.

Recognizing Common Illness Symptoms: Whether it's a fever, cough, headache, or skin rash, knowing how to identify the symptoms of common illnesses can guide your treatment decisions.

Assessing Dehydration and Malnutrition: In long-term survival situations, dehydration and malnutrition are common threats. Knowing the signs of both will help you take the necessary steps to prevent and treat these conditions.

c. Medication Administration

Basic knowledge of over-the-counter medications and their uses is also critical. While you may not have access to prescription drugs in a survival scenario, certain common medications can address a wide range of health issues:

Pain Management: Understanding the appropriate use of over-the-counter pain relievers such as acetaminophen, ibuprofen, or aspirin.

Antihistamines: Knowing when to use antihistamines for allergic reactions, especially in an environment where insect bites and stings are common.

Antibiotics: In some cases, you may have access to antibiotics. Understanding when and how to use them responsibly is essential to prevent resistance and ensure effectiveness.

d. Basic Wound Care

Managing injuries such as cuts, scrapes, abrasions, and blisters effectively is vital. Skills such as cleaning the wound, applying the appropriate dressing, and watching for signs of infection will help reduce the risk of complications.

3. Advanced Medical Skills for Preppers

While basic medical knowledge is crucial, there are several advanced medical skills that can be invaluable in more severe situations. These skills require more in-depth knowledge, practice, and often special tools or supplies. Acquiring advanced skills will make you more prepared for a wide range of emergency medical scenarios, particularly in extended or long-term survival situations.

a. Trauma and Emergency Care

In severe emergencies such as car accidents, natural disasters, or physical violence, trauma care becomes a key area of concern. Learning advanced trauma care techniques can significantly improve outcomes in critical situations. Some advanced skills include:

Advanced Wound Care: Techniques for dealing with deep lacerations, puncture wounds, and gunshot wounds.

Managing Internal Bleeding: Knowing how to recognize and respond to signs of internal bleeding, including applying pressure to control bleeding, and knowing when and how to use tourniquets.

Managing Fractures and Dislocations: Understanding the techniques for immobilizing fractures and dislocations, reducing fractures, and transporting patients safely.

b. Surgical Skills

In extreme cases where professional medical assistance is unavailable for days or weeks, you may need to perform basic surgery. Many prepper communities teach basic emergency surgery skills that could save lives in a critical situation. These skills include:

Laceration Repair: Knowing how to stitch up deep cuts and lacerations safely.

Abscess Drainage: Learning how to drain an abscess properly to avoid infection and relieve pressure.

Appendectomy and Other Minor Surgical Procedures: In some extreme situations, knowing how to perform an appendectomy or treat infections in the absence of antibiotics might be necessary.

c. Advanced Diagnostics and Monitoring

While basic diagnostics focus on identifying immediate health issues, advanced diagnostics include more sophisticated methods of determining the cause of illness or injury. This includes:

Blood Testing: Understanding how to use portable devices to test blood sugar, cholesterol, and other key indicators.

Ultrasound and Imaging: Some survivalists acquire portable ultrasound devices to monitor internal injuries, fluid buildup, or organ function.

d. Wilderness and Field Medicine

Many preppers recognize the importance of wilderness medicine for outdoor and off-grid survival. Wilderness medicine involves specialized skills that focus on treating injuries and illnesses in remote settings. These skills include:

Snake Bite Treatment: Knowing how to treat snake bites with first aid, suction devices, and applying a pressure bandage.

Hypothermia and Frostbite Management: Learning how to prevent and treat cold-related injuries, including frostbite and hypothermia.

Heat Exhaustion and Heat Stroke: Recognizing the symptoms of heat stress and knowing how to treat these conditions before they become life-threatening.

e. Understanding Medical Equipment

When preparing for long-term survival, preppers should learn to use and maintain medical equipment. Items like blood pressure cuffs, thermometers, pulse oximeters, and defibrillators can be critical in a survival scenario. Training in their use can provide a sense of confidence and security in treating both common and life-threatening issues.

4. How to Acquire These Skills

a. Take Classes and Certifications

One of the best ways to learn both basic and advanced medical skills is to attend formal classes. The American Red Cross and American Heart Association offer basic and advanced first aid and CPR certifications. For more advanced training, consider courses in wilderness first aid, trauma care, and even tactical combat casualty care (TCCC), which focuses on battlefield medicine.

b. Practice Regularly

Medical skills are perishable. Once you've learned a skill, it's essential to practice it regularly to maintain proficiency. This could involve doing mock drills, practicing wound care on mannequins, or working with a first aid buddy to simulate emergency situations.

c. Use Online Resources and Books

Many online resources and books can help you deepen your knowledge. Websites like Cochrane Collaboration, First Aid for Free, and MedPage Today offer free resources for learning medical skills. Books like The Survival Medicine Handbook by Joseph Alton and Where There Is No Doctor by David Werner are excellent references for prepping medical skills.

Basic and advanced medical skills are invaluable in a survival situation. Learning these skills will not only prepare you to respond effectively to common injuries and health issues but will also give you the confidence to handle life-threatening emergencies. From basic first aid to advanced trauma care and surgical procedures, the more you know, the better equipped you'll be to protect your loved ones and yourself. Investing in medical training is an essential part of your overall preparedness strategy, ensuring that you can manage emergencies and stay self-sufficient in the toughest of times.

Managing Chronic Conditions Without External Help

In a disaster scenario, whether it's a natural catastrophe, a grid-down situation, or civil unrest, the usual healthcare infrastructure may be unavailable. Hospitals and clinics might be overwhelmed, medical supplies may run low, and access to physicians could become scarce. For individuals with chronic conditions such as diabetes, hypertension, asthma, or heart disease, this situation can be especially challenging. Chronic conditions require ongoing management, and without proper care, they can quickly worsen, leading to serious complications.

This subchapter explores strategies for managing chronic conditions without external help, focusing on preparedness, self-management, and practical techniques that can be employed during prolonged emergencies. By understanding the impact of chronic conditions on your health and learning how to adapt and manage them with limited resources, you can ensure your survival and quality of life during emergencies.

1. Understanding Chronic Conditions and Their Impact in Crisis Situations

Chronic conditions are long-term health problems that require continuous or long-term care and management. Some of the most common chronic conditions include:

- Diabetes (Type 1 and Type 2)
- Hypertension (High Blood Pressure)
- Asthma and Chronic Obstructive Pulmonary Disease (COPD)
- Heart Disease
- Arthritis and other Musculoskeletal Disorders
- Epilepsy and other Neurological Disorders
- Kidney Disease
- Thyroid Disorders

During an emergency or survival situation, the management of these conditions can be disrupted, leading to severe consequences. For example, lack of insulin for a diabetic patient, or the unavailability of blood pressure medications for someone with hypertension, can result in life-threatening complications.

Managing chronic conditions without medical assistance requires both knowledge and preparation. The following strategies will help ensure that individuals with chronic conditions can navigate emergencies with minimal risk to their health.

2. Creating a Chronic Condition Management Plan

The key to managing chronic conditions without external help lies in preparation. The more proactive you are before a disaster strikes, the better positioned you will be to handle the situation. Creating a comprehensive management plan that addresses your chronic condition will help guide your actions and decisions when resources become scarce.

a. Assess and Document Your Condition

Start by understanding the specific nature of your chronic condition. Work with your healthcare provider before a crisis to get a clear understanding of:

The severity of your condition: How critical is it? Is it something that needs immediate intervention (e.g., insulin for diabetes or oxygen for COPD)?

Medications you rely on: List all medications, dosages, and frequency. Learn about potential substitutes or how to manage without them.

Lifestyle factors: What triggers your condition? Do you need specific types of food, exercise, or rest to manage it?

Possible complications: What are the emergency signs that your condition is worsening? When should you take extra precautions or seek help?

Ensure that this information is easily accessible, and store it in a place where you and family members can quickly reference it in an emergency.

b. Stockpile Critical Medications and Supplies

If possible, secure a sufficient supply of essential medications. Many chronic conditions, especially diabetes and heart disease, require ongoing medication. Consider:

Obtaining extra prescriptions: Work with your healthcare provider to arrange for additional refills. In some cases, long-term prescriptions can be filled in advance to create a stockpile.

Researching alternative medications: If you are unable to access your usual medication, it is important to know about alternatives. For example, if insulin is unavailable, ask your healthcare provider about potential substitutes such as oral diabetes medications or natural insulin-regulating strategies.

Supplies for managing the condition: In addition to medications, ensure you have the equipment necessary to monitor and manage your condition. For instance, diabetic patients should have blood glucose meters, test strips, syringes, and alcohol wipes. Those with asthma should have inhalers, nebulizers, and masks. Make sure these are regularly checked for expiration dates and stored in a safe, accessible location.

c. Identify Non-Pharmaceutical Management Strategies

While medication is often necessary for chronic conditions, there are several non-pharmaceutical management strategies that can be implemented during a crisis.

Dietary adjustments: Many chronic conditions can be managed or mitigated with changes to diet. For instance:

Diabetes: Focus on low glycemic foods to help stabilize blood sugar levels. Keep a supply of sugar-free snacks, natural sweeteners like stevia, and dried herbs like cinnamon, which can help regulate blood sugar.

Hypertension: Reducing sodium intake is essential for managing high blood pressure. Stock up on low-sodium foods and consider using salt alternatives like potassium chloride. Also, incorporating potassium-rich foods such as bananas and sweet potatoes can support heart health.

Heart disease: A heart-healthy diet includes whole grains, lean proteins, and healthy fats, such as those found in fish, nuts, and olive oil. Stock up on these foods to reduce the risk of cardiovascular events.

Herbal remedies: Some herbs can provide beneficial effects for managing chronic conditions. For instance:

Diabetes: Bitter melon and fenugreek are commonly used to lower blood sugar levels.

Hypertension: Hawthorn and garlic are well-regarded for their ability to manage blood pressure.

Asthma: Lobelia and eucalyptus can provide relief for asthma symptoms.

It is important to research any herbs or supplements to ensure they are safe and effective for your specific condition, especially in a crisis where you may not be able to get professional advice.

3. Lifestyle Adjustments for Self-Management

Maintaining a healthy lifestyle is crucial in managing chronic conditions, particularly when external help is unavailable. Here are some general lifestyle adjustments that can support your condition during an emergency:

a. Regular Monitoring

Self-monitoring is crucial in preventing complications from chronic conditions. Depending on your condition, this may involve:

- Blood glucose monitoring (for diabetes)
- Blood pressure readings (for hypertension)
- Oxygen saturation levels (for respiratory conditions)
- Tracking fluid intake and output (for kidney disease or heart failure)

Create a routine for checking these parameters and document them in a logbook. This can help you identify changes early and adjust your care plan accordingly.

b. Stress Management

Stress is a common trigger for worsening chronic conditions. Without access to professional help, managing stress becomes even more important. Techniques such as:

- Breathing exercises
- Meditation and mindfulness
- Gentle exercise or yoga
- Spending time outdoors in nature can help reduce stress and improve overall well-being.

c. Exercise and Mobility

Staying active is vital for managing chronic conditions like diabetes, heart disease, and arthritis. Regular movement improves circulation, strengthens the heart, and can even regulate blood sugar. Low-impact exercises like walking, swimming, or stretching can be beneficial, especially for older individuals or those with limited mobility.

4. Preparing for a Lack of External Help

In a situation where external help (such as doctors, hospitals, or pharmacies) is not available, being proactive about managing your condition is vital. Here are a few strategies:

a. Build a Support Network

It's essential to have a group of people you trust who can help if your health deteriorates. This can include family members, neighbors, or friends. Teach them about your condition, how to recognize emergencies, and how they can help manage your condition during a crisis.

b. Emergency Medical Supplies

Stock up on medical supplies for emergencies. This could include things like:

- Sterile bandages and gauze
- Antibiotic ointments
- Oxygen tanks and nebulizers (if required for respiratory conditions)
- Diabetic test strips and syringes

Having access to these supplies will allow you to better manage emergencies on your own.

c. Plan for Medication Substitutes

If you cannot get your prescribed medication during a disaster, having alternatives in place can help mitigate health risks. For example:

- **For diabetes:** Learn how to manage blood sugar through diet, exercise, and herbs if insulin becomes unavailable.
- **For hypertension:** Natural remedies, such as magnesium supplementation and potassium-rich foods, can help regulate blood pressure.

d. Recognize When Professional Help is Necessary

Some conditions cannot be managed without external help, especially in the case of life-threatening complications. For instance, a diabetic with a severe insulin shortage may need access to medical care.

Understanding when your condition is escalating beyond your ability to manage it is critical for long-term survival.

Managing chronic conditions without external help requires careful planning, a strong understanding of your condition, and the ability to make critical adjustments during an emergency. By stockpiling supplies, employing alternative treatments, and making necessary lifestyle adjustments, you can take control of your health and ensure you can survive and thrive in a crisis. While managing chronic conditions in isolation can be difficult, it is entirely possible with the right preparation, mindset, and knowledge.

Handling Psychological Stress and Trauma

In a disaster scenario, whether it's a natural catastrophe, civil unrest, or any other form of societal collapse, the physical survival of individuals often takes precedence. However, it's crucial not to overlook the psychological toll such events can have on both individuals and families. The emotional and mental health challenges during and after a crisis can be just as critical as physical survival skills. In fact, poor psychological health can exacerbate physical health problems, reduce decision-making abilities, and hinder effective response to emergencies.

This subchapter will explore the psychological impacts of emergencies, common mental health issues that arise in such situations, and strategies to manage stress and trauma effectively. By preparing yourself mentally and emotionally, you will increase your resilience and ability to navigate a crisis with strength and clarity.

1. Understanding Psychological Stress and Trauma in Crisis Situations

Psychological stress and trauma are common responses to emergencies, and their impact can vary depending on the severity of the event, personal coping mechanisms, and prior mental health history. Psychological stress occurs when an individual perceives a situation as threatening and struggles to cope with the pressure. Trauma, on the other hand, is a result of exposure to overwhelming or life-threatening events, which may lead to long-lasting emotional, psychological, and physical consequences.

During a disaster or prolonged emergency, individuals may experience a wide range of psychological reactions, such as:

Acute Stress: This is a natural, immediate response to a crisis. Symptoms may include anxiety, irritability, difficulty concentrating, and sleep disturbances. This is often short-term and can resolve as the situation stabilizes.

Post-Traumatic Stress Disorder (PTSD): Prolonged exposure to traumatic events can lead to PTSD, a more severe form of stress. Symptoms may include flashbacks, nightmares, emotional numbness, hypervigilance, and avoidance behaviors. PTSD can develop days, months, or even years after the event.

Depression and Anxiety: Living in stressful conditions or experiencing the loss of a loved one, home, or resources can trigger depression or chronic anxiety. Feelings of hopelessness, sadness, and an inability to cope with daily life are common.

Survivor's Guilt: After a disaster, survivors may feel guilty for having made it through when others did not. This can contribute to depression, anxiety, and emotional distress.

Cognitive Impairment: High levels of psychological stress can impair cognitive functions like memory, decision-making, and problem-solving, which are critical in survival situations.

Understanding these responses is key to managing your psychological health during a crisis. Recognizing the signs of stress and trauma in yourself and others will allow for early intervention and the adoption of coping strategies to mitigate the mental health impact.

2. Strategies for Managing Psychological Stress

While mental health challenges can be overwhelming, there are numerous strategies that can help mitigate the psychological impact of stressful situations. By preparing ahead of time, you can develop resilience and ensure that you have tools in place to handle emotional distress.

a. Building Emotional Resilience

Emotional resilience is the ability to bounce back from adversity and cope with difficult situations. In a survival scenario, being emotionally resilient is crucial for making clear decisions, maintaining relationships, and staying physically healthy. Here are ways to build emotional resilience:

- **Develop a Support System:** A strong social support network is one of the most important factors in emotional resilience. Having family members, friends, or even neighbors you can rely on during a crisis will reduce feelings of isolation and fear.
- **Practice Mindfulness and Relaxation Techniques:** Mindfulness helps individuals stay grounded in the present moment, preventing them from getting overwhelmed by future uncertainties. Techniques like deep breathing, meditation, or progressive muscle relaxation can help lower stress levels and keep you focused.
- **Adopt a Positive Mindset:** In challenging situations, it is easy to feel helpless, but maintaining a positive attitude is essential. Even in the darkest times, focusing on what you can control, and taking small, practical steps forward can help build hope.
- **Develop Problem-Solving Skills:** By preparing for crises in advance and practicing your response strategies, you will be more confident in your ability to handle the situation, which reduces psychological stress.

b. Managing Anxiety and Fear

During a crisis, anxiety and fear can become paralyzing, leading to poor decision-making and inability to take action. Here are strategies for managing anxiety and fear:

Controlled Breathing: Slow, deep breaths can help calm your nervous system during moments of panic or anxiety. Try the 4-7-8 technique: breathe in for 4 seconds, hold for 7 seconds, and exhale for 8 seconds.

Focus on What You Can Control: Anxiety often stems from fear of the unknown. By focusing on immediate, manageable tasks—such as securing food, water, and shelter—you can regain a sense of control and calm your mind.

Stay Connected: Anxiety can increase when individuals feel isolated or cut off from others. Keeping in touch with family, friends, or your community provides a sense of support and stability.

Limit Exposure to Negative News: During a crisis, constant media consumption can exacerbate fear and anxiety. Limit your exposure to news coverage and focus on practical, actionable steps.

3. Trauma-Informed Care for Yourself and Others

In crisis situations, it is important to recognize the signs of trauma and provide care not only for others but also for yourself. Survivors of traumatic events can experience a range of emotions, from shock and disbelief to anger and despair. Providing trauma-informed care means understanding and addressing the effects of trauma, both immediately and in the long-term. This is especially critical if you're responsible for the well-being of others, such as children, elderly family members, or individuals with pre-existing mental health conditions.

a. Creating a Safe Environment

Safety is one of the most important factors in managing trauma. Survivors need to feel physically and emotionally safe. This can include:

Ensuring physical shelter: Creating a secure, safe home environment can help reduce anxiety.

Providing emotional support: Listen to people's concerns without judgment. Create a space where individuals feel comfortable sharing their feelings.

Maintaining routine: As much as possible, maintain a sense of normalcy. Routine can provide comfort and stability, especially in highly uncertain situations.

b. Active Listening and Empathy

When helping others cope with trauma, practice active listening. Give people the space to express their feelings without interrupting or offering unsolicited advice. Validation of their emotions is critical to helping them process their experience.

c. Seeking Professional Support

If possible, seek professional mental health support. Even during a crisis, there may be telehealth services or other crisis counseling resources available. If not, having a basic understanding of psychological first aid and other trauma-informed techniques can help you assist others.

4. The Role of Sleep and Nutrition in Mental Health

Physical well-being is directly linked to mental health. In emergency situations, getting enough sleep and maintaining proper nutrition are two critical components of managing stress.

a. Sleep

Sleep is essential for psychological resilience. It allows the brain to repair itself, process emotions, and regulate mood. In stressful situations, sleep may be disrupted, leading to cognitive impairments and increased emotional instability. Here are a few tips to ensure better sleep during a crisis:

Create a Sleep-Conducive Environment: Make sure your sleeping space is quiet, dark, and comfortable. Use earplugs or white noise if needed.

Establish a Sleep Routine: Try to maintain a regular sleep schedule even in chaotic circumstances. This helps your body stay in a rhythm and improves the quality of sleep.

b. Nutrition

Proper nutrition is crucial for maintaining mental clarity and emotional stability. During a crisis, people may have limited access to food, which can exacerbate stress. Stock up on nutrient-dense, easy-to-prepare foods such as:

- Omega-3-rich foods (like fish, walnuts, and flaxseeds) to support brain function
- Complex carbohydrates (whole grains and vegetables) to stabilize blood sugar levels and enhance mood
- Foods high in magnesium and zinc (such as leafy greens and pumpkin seeds) to reduce anxiety
- Hydration is also important for mental clarity, so ensure access to clean drinking water.

5. Long-Term Strategies for Mental Health and Recovery

After a crisis or trauma, long-term recovery is necessary. It may take weeks, months, or even years to process traumatic experiences fully. Survivors may need ongoing mental health support, but there are also proactive steps they can take:

- **Maintain connections:** Stay in contact with friends, family, and support groups. Social support can prevent isolation and provide emotional comfort.
- **Engage in meaningful activities:** Rebuilding a sense of purpose and engaging in activities that provide fulfillment (such as hobbies or work) can promote healing.
- **Practice self-care:** Take time for self-care, whether that's taking walks, practicing mindfulness, or engaging in creative endeavors. Healing requires giving yourself permission to rest and rejuvenate.

Handling psychological stress and trauma is essential in ensuring your survival and well-being during a disaster or emergency. By understanding the mental health challenges that can arise in these situations, you can adopt strategies to build emotional resilience, manage anxiety and stress, and support both yourself and others. With proper preparation, mental clarity, and support, you can navigate a crisis with strength and stability, ensuring that your mental health remains as resilient as your physical survival skills.

11. STAYING CONNECTED AND INFORMED

Accessing Reliable News and Alerts

In times of crisis or emergency, one of the most critical components of your survival strategy is having access to reliable, real-time information. News and alerts help you stay informed, make informed decisions, and respond to a crisis with accuracy and preparedness. In a world where misinformation spreads easily, especially during emergencies, being able to distinguish between credible sources and unreliable ones can make all the difference in ensuring your safety and the safety of your loved ones.

This subchapter will explore why accessing reliable news and alerts is vital during emergencies, how to find trustworthy sources, and how to set up systems that ensure you receive the most up-to-date and accurate information possible.

1. The Importance of Accessing Reliable Information During a Crisis

In a disaster or emergency scenario—whether it's a natural disaster, civil unrest, or a public health crisis—having timely and accurate information is crucial. The consequences of receiving unreliable or delayed news can range from missed evacuation orders to unsafe decision-making or even increased panic. Here are several reasons why accessing reliable news and alerts is critical:

a. Timely Decision-Making

When disaster strikes, the window for effective decision-making can be very narrow. For example, in a wildfire, storm, or civil unrest scenario, knowing exactly where the danger is, where safe zones are, and whether you need to evacuate or shelter in place can save lives. Access to reliable, real-time information allows you to make critical decisions quickly.

b. Accurate Risk Assessment

News and alerts help you assess the severity of the crisis. If you rely on inaccurate information or rumors, you might misjudge the situation. For instance, during a pandemic, inaccurate news can lead to unnecessary panic or, conversely, a failure to take adequate precautions. Access to verified news sources allows you to understand the risk and adjust your response accordingly.

c. Avoiding Panic and Misinformation

Emergencies often bring about fear and confusion, which can lead to the spread of misinformation. In such situations, it's easy to be overwhelmed by rumors or sensationalized news. Reliable news and alerts help cut through the noise, preventing you from making decisions based on unfounded rumors or false information.

d. Staying Informed About Relief Efforts and Resources

During a crisis, knowing where to find relief resources (such as emergency shelters, food distribution centers, medical aid, and evacuation routes) can be lifesaving. Reliable news sources will often provide updates on where to find such resources and how to access them, helping you and your family stay safe and secure.

2. Types of Alerts and News Sources to Rely On

In an age of digital connectivity, there are numerous ways to access news and alerts. However, not all sources are created equal, and it's important to choose reliable, accurate, and timely options. Below are some of the best types of news and alert systems to consider:

a. National and Local Government Alerts

Government agencies are typically among the most reliable sources of information during a crisis. Many countries and local governments have emergency alert systems designed to send real-time notifications to citizens about public safety, emergencies, and health updates. These alerts are designed to keep you informed about evacuations, shelter-in-place orders, weather warnings, and more.

Wireless Emergency Alerts (WEA): These alerts are sent directly to your mobile phone and can include notifications about weather events, local emergencies, or public safety issues. Check if your mobile device is set to receive these alerts.

Federal Emergency Management Agency (FEMA) Alerts (USA): FEMA offers alerts through its Integrated Public Alert and Warning System (IPAWS), which provides critical emergency alerts on a local, state, and national level.

Local Emergency Management Services (EMS): Local EMS departments, fire stations, or police can send out alerts through text messages, emails, or social media during local disasters.

b. Trusted News Outlets

During a crisis, turning to established news organizations can provide accurate, in-depth coverage of the situation. When looking for reliable news sources, aim for outlets that have a history of credibility, adherence to journalistic standards, and a commitment to fact-based reporting. Some well-regarded news outlets include:

BBC (British Broadcasting Corporation): Known for its global coverage of major events, BBC provides updates on emergencies worldwide, with an emphasis on clear, unbiased reporting.

Reuters and Associated Press (AP): Both organizations are renowned for providing factual, up-to-date news from trusted reporters on the ground.

National Public Radio (NPR): Known for offering detailed coverage, NPR provides both news reports and expert commentary, focusing on national and international emergencies.

Your Local News Stations: Local news outlets often provide the most up-to-date information for your area, including coverage of localized natural disasters, power outages, and civil unrest.

c. Social Media for Real-Time Updates

While social media is a source of much misinformation, it can also be an invaluable tool during an emergency when used correctly. Following trusted sources on social media can help you receive quick, live updates on an unfolding situation. It's crucial to follow accounts of:

- Government Agencies (e.g., local police, fire departments, and disaster management organizations)
- Verified News Outlets (many news stations maintain active Twitter accounts or Facebook pages with live updates)
- Trusted Nonprofits and Relief Organizations (e.g., Red Cross, Doctors Without Borders)

For example, platforms like Twitter or Telegram often provide real-time updates on breaking news, natural disasters, and emergencies. Just be cautious about the source of the information to avoid falling for rumors or unverified claims.

d. NOAA Weather Radio (USA)

In the United States, NOAA (National Oceanic and Atmospheric Administration) Weather Radio broadcasts continuous broadcasts of official Weather Service warnings, watches, and other emergency information. These broadcasts are available 24/7 and can provide reliable weather updates during severe weather events like hurricanes, tornadoes, or floods. NOAA Weather Radio is a must-have item for preppers who live in areas prone to natural disasters.

e. Apps for Emergency Alerts

Many modern apps are designed to send notifications about local emergencies directly to your smartphone. These apps can be customized to your location, ensuring you receive alerts that are most relevant to you. Some popular emergency alert apps include:

- **Red Cross Emergency App:** Offers real-time alerts and provides safety tips, checklists, and emergency preparedness guides.
- **Disaster Alert App:** Sends real-time global disaster alerts, including earthquakes, wildfires, tsunamis, and other natural disasters.
- **Storm Shield:** Provides weather warnings and alerts for hurricanes, tornadoes, and severe storms.

3. How to Set Up a Reliable Information Network

To maximize the effectiveness of your emergency response, it's essential to establish a robust information network well before a disaster strikes. Here's how you can build and optimize your access to reliable news and alerts:

a. Subscribe to Local Alert Systems

Sign up for emergency alert services provided by local government agencies. Many local governments and utilities offer text message alerts or email notifications about power outages, severe weather, and evacuations.

b. Diversify Your Sources

Relying on a single source of information during a crisis can leave you vulnerable to misinformation or gaps in coverage. Use a combination of government alerts, news outlets, and social media updates to ensure that you receive the most comprehensive and accurate picture of the situation.

c. Equip Yourself with Multiple Devices

If possible, access information through various devices. For example, a NOAA Weather Radio can provide alerts when the power is out and your phone is dead. Ensure that your smartphone, tablet, and any battery-operated radios or devices are charged and ready to receive alerts.

d. Set Up Push Notifications

Activate push notifications for relevant news apps and alert systems. This ensures that you are immediately notified when a crisis or emergency breaks. Be sure to prioritize critical alerts, like weather warnings and evacuation notices.

e. Verify Information

Before acting on any information, especially if it comes from social media, verify it through official sources such as government agencies or trusted news outlets. Cross-reference details to avoid reacting to rumors or misinformation.

4. Preparing for Communication Blackouts

In some cases, communication channels may become overloaded or shut down due to the crisis. In these situations, it's vital to have backup methods of communication, such as:

- **Two-Way Radios:** For localized emergencies, two-way radios can provide an alternative to mobile phones.
- **Ham Radio:** In extreme cases, when cell towers are down, ham radio operators often play a key role in emergency communications.
- **Satellite Phones:** These phones can function in areas where traditional cell networks are unavailable.

Accessing reliable news and alerts is a fundamental aspect of surviving and thriving in any emergency. The ability to make informed decisions, stay ahead of threats, and understand available resources is critical to your survival and long-term success in a crisis. By diversifying your sources of information, setting up reliable alert systems, and learning how to discern fact from fiction, you can ensure that you are well-prepared to respond to any emergency with confidence and clarity.

Using Backup Communication Tools and Radios

In a crisis or emergency situation, communication is paramount to ensuring the safety of you and your loved ones. While cell phones and the internet are often the go-to communication tools in day-to-day life, they can become unreliable or unavailable during a disaster, such as a power outage, natural disaster, or civil unrest. This is where backup communication tools, including radios and satellite phones, play a vital role in maintaining contact when traditional methods fail.

In this subchapter, we will explore the importance of backup communication, the types of tools available, how they work, and how to set up an effective communication system during emergencies.

1. Why Backup Communication Tools Are Essential

During a crisis, communication becomes crucial for several reasons:

Coordinating evacuations: You may need to reach family members to coordinate evacuations or shelter-in-place instructions.

Receiving emergency updates: During events like hurricanes, wildfires, or civil unrest, staying informed about the situation is critical for safety.

Sharing critical information: Whether it's reporting an injury, a fire, or a structural issue, being able to send information or ask for help can make the difference between life and death.

Navigating power outages: Many communication systems, like mobile phones and the internet, rely on electricity. In the event of a widespread power outage, having a backup system is essential.

Backup communication tools ensure that you are not reliant on a single system that may fail when you need it the most. By having multiple communication methods available, you increase your chances of staying connected during a disaster.

2. Types of Backup Communication Tools

Several types of communication tools are essential for preppers and those looking to prepare for emergencies. These tools can help bridge the gap when traditional communication methods become unavailable.

a. Two-Way Radios (Walkie-Talkies)

Two-way radios, also known as walkie-talkies, are one of the most widely used communication tools in emergency situations. They allow for direct, short-range communication without relying on cellular networks or the internet.

How They Work: Two-way radios work by transmitting and receiving radio waves, and they can be used over relatively short distances, typically from a few hundred feet up to several miles, depending on the model and terrain. The range is influenced by factors like terrain, weather, and obstacles.

Benefits: Walkie-talkies are portable, simple to use, and don't rely on the internet or cell towers. They are ideal for keeping in touch with family members within a limited area, such as during an evacuation or while navigating an emergency zone.

Popular Models: Some well-regarded models of two-way radios include the Motorola T600 Talkabout (with a range of up to 35 miles in open terrain) and the Baofeng UV-5R, which is a popular choice for ham radio operators and emergency preparedness.

Tips for Use:

Always check the communication range in your area before relying on them for emergencies.

Invest in rechargeable batteries and keep spares to ensure functionality.

Pre-program radio frequencies that might be relevant during emergencies (like local emergency channels or ham radio frequencies).

b. Hand-Crank Radios

Hand-crank radios, also known as emergency radios, are another key tool in an emergency communication plan. These radios can be powered through a hand crank, solar power, or batteries, making them ideal when power outages occur.

How They Work: Hand-crank radios convert mechanical energy from cranking into electrical energy, allowing you to power the device without needing electricity or batteries. Many models also include a built-in flashlight and a USB port to charge mobile devices.

Benefits: These radios provide access to weather alerts, emergency broadcasts, and AM/FM radio stations. Many models also feature NOAA (National Oceanic and Atmospheric Administration) Weather Radio, which broadcasts continuous weather information, including urgent emergency weather alerts, making them an essential tool for staying informed during a natural disaster.

Popular Models: Well-known hand-crank radio models include the Eton American Red Cross FRX3+ and the Kaito KA500. Both offer NOAA weather alerts and multiple power options, including hand crank, solar panel, and USB charging.

Tips for Use:

Test the radio periodically to ensure it's functional, especially if you've stored it for a while.

If using a hand-crank model, avoid over-cranking the device to prevent damage.

Consider keeping spare batteries or a solar charger to supplement the hand-crank method, especially for extended use.

c. Satellite Phones

Satellite phones are an excellent backup communication tool for areas with no cellular network coverage, such as remote locations or when communications infrastructure has been damaged or destroyed during a disaster.

How They Work: Unlike traditional mobile phones, which rely on cell towers, satellite phones connect directly to satellites orbiting the Earth, allowing you to make and receive calls from almost any location on the planet.

Benefits: Satellite phones are reliable for staying in touch with loved ones or emergency services during crises where traditional communication methods fail. They provide coverage even in remote areas or when natural disasters damage local cell towers.

Popular Models: The Iridium 9575 Extreme and the Inmarsat IsatPhone 2 are two of the most popular satellite phone models, providing reliable global coverage.

Tips for Use:

Be sure to test your satellite phone before an emergency to ensure that you understand how to use it and that it works.

Satellite phones require a clear line of sight to the sky, so you may need to move to an open area to make or receive calls.

d. Ham Radio (Amateur Radio)

Ham radios, or amateur radios, are used by licensed operators to communicate over long distances. While more technical than walkie-talkies, ham radios are one of the most reliable backup communication tools in an emergency.

How They Work: Ham radios transmit over both short and long distances using various frequencies and communication modes. They rely on operators who hold amateur radio licenses, which are required to operate legally.

Benefits: Ham radios provide access to a global network of amateur radio operators, many of whom monitor emergency frequencies and provide valuable assistance during crises. Ham radios can be used to send and receive messages when no other communication systems are available, making them a valuable tool in the preparedness toolkit.

Popular Models: The Yaesu FT-60R and Icom IC-7300 are popular models for both beginners and experienced operators. Ham radios require a license to operate, so make sure to complete the necessary training before use.

Tips for Use:

- Obtain your amateur radio license through the appropriate training programs and exams.
- Learn how to operate your ham radio and familiarize yourself with emergency communication protocols.
- Keep spare batteries or a solar-powered charging system on hand to ensure continuous operation.

e. Messaging Devices

In some cases, it may not be necessary to speak directly to someone but rather send a short, reliable message. Devices like satellite messengers and personal locator beacons (PLBs) can be extremely useful in emergencies where phone networks are unavailable.

How They Work: Satellite messengers like the Garmin inReach or the SPOT Gen4 send short text messages via satellite, allowing you to communicate even when there is no cell service. These devices are popular among outdoor adventurers and preppers who spend time in remote areas.

Benefits: They allow you to send critical messages, such as requesting help or providing your coordinates, without requiring a phone or internet service.

Popular Models: The Garmin inReach Explorer+ and the SPOT Gen4 Satellite Messenger are popular choices that provide two-way messaging and GPS location tracking.

3. Setting Up an Effective Communication System

To ensure that you and your family stay connected during an emergency, it's essential to have a well-organized communication system in place. Here are a few steps to help you set up a system that works:

a. Create a Communication Plan

Your communication plan should include details on how to reach each family member, what tools to use, and backup options if primary methods fail. Ensure everyone in your household understands how to use the backup communication tools you've chosen.

Designate meeting points: In case you can't reach someone by phone, have designated meeting spots where you and your loved ones can gather if safe to do so.

Practice using the tools: Regularly practice using two-way radios, satellite phones, or ham radios to ensure everyone is familiar with the devices and can use them in an emergency.

b. Keep Tools Accessible and Functional

Backup communication tools should be easily accessible and ready to use at a moment's notice. Keep these tools charged, maintain spare batteries, and store them in a waterproof, easy-to-reach location.

c. Consider Power and Battery Options

Many backup communication tools require power, and during an emergency, power may not be readily available. Consider investing in solar-powered chargers, extra batteries, and portable battery banks to keep your tools operational.

d. Stay Informed About Frequency Changes

For devices like ham radios, be sure to stay informed about frequency changes, especially in the case of government or emergency services announcing specific channels for distress communication. Joining a local amateur radio club can also help you stay in touch with others who may be able to provide useful information during an emergency.

Coordinating with Neighbors and Networks

Effective coordination with neighbors and community networks is essential for enhancing resilience and ensuring safety during emergencies. By fostering strong relationships and collaborative preparedness efforts, communities can better respond to and recover from disasters. This subchapter explores the importance of neighborly coordination, strategies for building robust community networks, and practical steps to implement these initiatives.

1. The Importance of Neighborly Coordination

In times of crisis, communities with established networks are better equipped to handle emergencies. Research indicates that individuals who know and trust their neighbors report higher rates of health and well-being compared to those who do not. citeturn0search11

Coordinating with neighbors offers several advantages:

- **Resource Sharing:** Pooling resources such as tools, food, and medical supplies can be more efficient than individual stockpiling.
- **Information Dissemination:** Neighbors can share critical information about local hazards, evacuation routes, and emergency services.
- **Mutual Aid:** Assisting each other with tasks like childcare, pet care, or medical needs can alleviate stress during emergencies.
- **Community Resilience:** Strong community ties contribute to a quicker and more effective recovery post-disaster.

2. Building Robust Community Networks

Establishing a resilient community network involves proactive planning and engagement. Consider the following steps:

- **Assess Community Needs and Resources:** Identify local hazards, available resources, and the skills of community members. This assessment helps in understanding the community's vulnerabilities and strengths.
- **Develop a Community Emergency Plan:** Create a comprehensive plan that outlines roles, responsibilities, and communication strategies during emergencies. Regularly review and update this plan to ensure its effectiveness.
- **Establish Communication Channels:** Set up reliable communication methods, such as group messaging apps, community bulletin boards, or local radio stations, to disseminate information quickly.
- **Organize Training and Drills:** Conduct regular training sessions and emergency drills to prepare community members for various scenarios. This practice enhances readiness and builds confidence.

- **Foster Inclusivity:** Ensure that all community members, including vulnerable populations, are included in planning and preparedness activities. This inclusivity strengthens the overall resilience of the community.

3. Practical Steps for Neighborly Coordination

To implement effective coordination with neighbors, consider the following actions:

Host Community Meetings: Organize regular gatherings to discuss preparedness plans, share information, and build relationships.

Create a Neighborhood Directory: Develop a contact list of neighbors, noting special skills or resources (e.g., medical expertise, generators) that can be valuable during emergencies.

Establish a Mutual Aid Agreement: Form agreements outlining how neighbors will assist each other during emergencies, including sharing resources and providing support.

Utilize Technology: Leverage social media platforms and community apps to facilitate communication and coordination. Platforms like Nextdoor can be effective for local information sharing.

Engage in Community Projects: Participate in local initiatives such as neighborhood clean-ups, gardening projects, or skill-sharing workshops to strengthen community bonds.

4. Case Studies and Examples

Communities worldwide have successfully implemented neighborly coordination strategies:

Townsville, Australia: In response to past disasters, Townsville developed a disaster recovery directory to provide centralized access to support and government directives, enhancing community resilience.

Rural Women's Health Project (RWHP), USA: After Hurricane Helene, RWHP coordinated evacuation messages, distributed essential supplies, and conducted welfare checks, targeting Spanish-speaking immigrant communities who may fear government shelters.

5. Overcoming Challenges

While coordinating with neighbors offers numerous benefits, challenges may arise:

- **Diverse Needs:** Addressing the varied needs of community members, including those with disabilities or language barriers, requires thoughtful planning.
- **Resource Limitations:** Communities may face constraints in resources and funding, necessitating creative solutions and partnerships.
- **Sustaining Engagement:** Maintaining active participation in preparedness activities can be challenging; regular engagement and clear communication are key to sustaining interest.

12. FIRE SAFETY AND HOME MANAGEMENT

Preventing House Fires During Crisis Scenarios

During crisis situations—such as natural disasters, power outages, or civil unrest—the risk of house fires can increase due to compromised infrastructure, heightened stress, and altered routines. Understanding and implementing fire prevention strategies during these times is crucial for safeguarding lives and property.

1. Understanding Increased Fire Risks During Crises

In crisis scenarios, several factors can elevate the risk of house fires:

Compromised Utilities: Power outages may lead individuals to use alternative heating or cooking methods, such as portable generators or open flames, which can be hazardous if not used properly.

Disrupted Emergency Services: Emergency response times may be delayed, reducing the effectiveness of firefighting efforts.

Stress and Distraction: The stress of a crisis can lead to lapses in attention, increasing the likelihood of accidents like unattended cooking or improper use of heating devices.

2. Fire Prevention Strategies During Crises

To mitigate fire risks during crises, consider the following measures:

Safe Use of Alternative Heating Sources:

Portable Generators: Never operate portable generators indoors or in enclosed spaces due to the risk of carbon monoxide poisoning. Always refuel generators outdoors and in well-ventilated areas.

Open Flames: Avoid using open flames for heating. If necessary, use flameless alternatives like battery-operated heaters designed for indoor use.

Safe Cooking Practices:

Unattended Cooking: Never leave cooking appliances unattended, especially when using alternative cooking methods. If you must leave the area, turn off the appliance.

Use of Stoves and Ovens: Do not use stoves or ovens to heat your home. These appliances are designed for cooking, not heating, and can pose fire hazards if misused. citeturn0search0

Electrical Safety:

Avoid Overloading Circuits: Do not overload electrical outlets or power strips, as this can lead to overheating and potential fires.

Inspect Electrical Equipment: Before use, check all electrical equipment for damage. Do not use damaged cords or appliances.

Fire Safety Equipment:

Smoke Alarms: Ensure smoke alarms are installed on every level of your home, including inside and outside bedrooms. Test alarms monthly and replace batteries at least once a year. citeturn0search2

Fire Extinguishers: Keep fire extinguishers accessible in key areas, such as the kitchen and near heating sources. Ensure all household members know how to use them.

Emergency Preparedness:

Fire Escape Plan: Develop and practice a fire escape plan with all household members. Include two ways out of every room and designate a meeting spot outside. citeturn0search2

Stay Informed: Monitor local news and official channels for updates on the crisis and fire safety advisories.

3. Special Considerations During Specific Crises

Natural Disasters:

Earthquakes: Secure heavy furniture and appliances to prevent them from tipping over and causing fires.

Floods: Avoid using electrical appliances if water has entered your home to prevent electrical hazards.

Power Outages:

Lighting: Use flashlights or battery-operated lanterns instead of candles to reduce fire risks.

Refrigeration: Keep refrigerator and freezer doors closed to maintain food safety during outages.

Civil Unrest:

Home Security: Secure all entry points to prevent unauthorized access, which could lead to accidental fires.

Emergency Supplies: Maintain a supply of non-perishable food, water, and necessary medications to reduce the need for cooking during periods of unrest.

4. Community Resources and Support

During crises, local fire departments and emergency services may offer resources and support:

Fire Safety Workshops: Participate in community workshops to learn about fire prevention and emergency response.

Emergency Shelters: Know the locations of local shelters and their fire safety protocols.

Assistance Programs: Some communities offer programs to assist with fire safety equipment installation, such as smoke alarms and fire extinguishers.

Preventing house fires during crisis scenarios requires proactive planning, adherence to safety protocols, and community engagement. By implementing these strategies, individuals and families can significantly reduce the risk of fire-related incidents during challenging times.

Cooking and Heating Alternatives Without Power

When the power goes out, it can disrupt daily routines and create significant challenges for survival. However, with proper preparation and the use of alternative cooking and heating methods, you can continue to meet your essential needs without relying on electricity. This chapter explores practical, safe, and efficient ways to cook and heat your home during power outages, focusing on long-term sustainability and ease of use.

1. Alternative Cooking Methods Without Power

In the absence of electricity, there are numerous ways to cook meals that don't require a stove or microwave. Here are some of the most effective cooking methods:

A. Propane Stoves and Grills

Propane stoves and grills are among the most popular and efficient cooking alternatives during power outages. Portable propane stoves are compact and easy to store, making them an ideal choice for emergency situations. These stoves typically use small canisters of propane, which are readily available and can provide a reliable cooking source for an extended period.

Advantages:

- Propane burns cleanly and efficiently, producing minimal smoke or ash.
- The fuel is widely available, and portable stoves are easy to use and maintain.
- Propane stoves can be used both indoors (in well-ventilated areas) and outdoors.

Considerations:

- Always ensure proper ventilation when using propane indoors to avoid the buildup of carbon monoxide.
- Stock up on propane canisters in advance, as supplies may be limited during an emergency.

B. Campfires and Fire Pits

A campfire or fire pit is a traditional yet effective method for cooking food and providing warmth. In a survival situation, this method is particularly valuable, as it requires minimal equipment and can be adapted to various types of cooking (e.g., boiling, roasting, baking).

Advantages:

- Firewood is abundant in many areas, and you can cook a wide variety of meals over an open flame.
- Campfires provide heat, light, and a cooking surface all at once.
- The flexibility of cooking methods, such as grilling, roasting, or making stews.

Considerations:

- Building a campfire requires some skill to ensure it is safe and efficient.
- Fire safety is critical, especially in dry conditions where wildfires can quickly spread.
- Outdoor use is ideal, but it's possible to cook indoors with a fire pit if you have the proper ventilation.

C. Solar Ovens

Solar ovens use the power of the sun to cook food, making them an environmentally friendly and energy-efficient choice for long-term survival. Solar cookers are especially useful in sunny climates and require no fuel other than sunlight.

Advantages:

- They are entirely fuel-free and have minimal operating costs.
- Solar ovens are safe to use and produce no smoke or toxic fumes.
- They can cook food slowly over time, allowing you to prepare meals without constant attention.

Considerations:

- Solar ovens are weather-dependent, requiring clear skies and direct sunlight for effective cooking.
- They are generally slower than traditional cooking methods, so planning ahead is necessary.

D. Wood-Burning Stoves

Wood-burning stoves, or "rocket stoves," are an excellent alternative to electric or gas stoves. These stoves use small amounts of wood to produce significant heat, making them efficient for both cooking and heating.

Advantages:

- Very efficient in terms of fuel usage; small amounts of wood can generate a lot of heat.
- Can be used both indoors and outdoors, depending on the design.
- Suitable for long-term use when paired with a steady supply of firewood.

Considerations:

- Requires a supply of dry, seasoned wood for consistent cooking and heating.
- Indoor use requires proper ventilation to avoid carbon monoxide buildup.

- Can be bulky, so portable versions are preferred for emergencies.

E. Alcohol Stoves

Alcohol stoves are compact, lightweight cooking devices that use alcohol-based fuels (usually ethanol or methanol). These stoves are commonly used by campers and hikers and are a reliable source of cooking heat during a power outage.

Advantages:

- Very portable and easy to store.
- Alcohol burns cleanly, producing minimal fumes and smoke.
- Alcohol can often be found in household products such as rubbing alcohol or can be purchased as fuel.

Considerations:

- Alcohol stoves are best suited for light cooking (e.g., boiling water, heating food) rather than heavy-duty meals.
- Fuel can be more expensive than propane, and stocking up in advance is important.

F. Butane Stoves

Butane stoves operate similarly to propane stoves but use butane canisters as fuel. These stoves are popular for camping and outdoor cooking and can be adapted for use during power outages.

Advantages:

- Butane stoves are compact, efficient, and easy to use.
- They are lightweight and portable, ideal for emergency situations.
- The fuel is easy to store and generally cheaper than propane.

Considerations:

- Butane is less effective in very cold temperatures, so it may not be ideal for winter use.
- Always use butane stoves in well-ventilated areas to avoid carbon monoxide buildup.

2. Heating Alternatives Without Power

Maintaining warmth during a power outage, especially in winter, is essential for comfort and survival. Several alternative heating methods can provide warmth and help you endure cold temperatures.

A. Propane Heaters

Portable propane heaters are a reliable option for heating in emergencies. These heaters are designed to be used indoors and produce significant heat without requiring electricity.

Advantages:

- Propane heaters are portable, easy to operate, and can heat a room effectively.
- Some models come with safety features such as automatic shutoff and oxygen depletion sensors.

Considerations:

- Adequate ventilation is critical to avoid carbon monoxide poisoning.
- Always follow manufacturer guidelines for indoor use and never leave a propane heater unattended.

B. Wood Stoves

Wood stoves are a traditional and highly efficient method of heating during power outages. When combined with a fireplace, a wood stove can provide heat and a cooking surface.

Advantages:

- Wood stoves can heat an entire house if the right model is used.
- They use readily available firewood, which can be gathered and stored for long-term use.
- Can be used to heat water or cook food, making them a versatile option.

Considerations:

- Requires regular cleaning and maintenance to ensure proper functioning.
- Indoor use must include a chimney or flue system to vent smoke safely.

C. Kerosene Heaters

Kerosene heaters are portable devices that burn kerosene to provide warmth. These heaters are commonly used for emergency heating and can quickly warm a small area.

Advantages:

- Kerosene heaters are highly effective in warming rooms.
- They are affordable and can be used in various situations, both indoors and outdoors.

Considerations:

- Like propane, kerosene heaters require ventilation to avoid carbon monoxide buildup.
- Fuel storage can be problematic, and kerosene may have an unpleasant odor.

D. Electric Blankets and Heated Pads

If your power outage is brief and you have access to backup batteries or generators, electric blankets or heated pads can be an effective way to stay warm without using much energy.

Advantages:

- Provides direct, personal warmth, making it ideal for sleeping or resting.
- Compact and easy to store.

Considerations:

- Requires a power source such as a battery or generator.
- Can be less effective during extended power outages if no backup power is available.

Surviving without power during a crisis requires creative thinking and preparation. Alternative cooking and heating methods provide crucial solutions for ensuring you have the means to survive and thrive during power outages. By stocking up on the appropriate equipment, fuel, and knowledge, you can face the challenge of a power outage with confidence and ease. Whether you opt for portable stoves, campfires, or wood-burning heaters, the key is preparation—ensuring that you have reliable, safe, and sustainable alternatives ready to meet your needs during a crisis.

Emergency Fire Suppression Techniques

Fire is a critical threat during an emergency or disaster scenario. Whether caused by natural disasters, accidents, or civil unrest, fires can spread quickly, destroy property, and claim lives. As a result, knowing effective fire suppression techniques is vital for survival and protection in crisis situations. This chapter covers essential fire suppression methods, from basic tools to advanced techniques, for dealing with fires during emergencies.

1. Understanding Fire Behavior

Before diving into specific fire suppression techniques, it's important to understand how fires behave. Fires require three elements to sustain themselves: heat, fuel, and oxygen. This is known as the "fire triangle." To effectively suppress a fire, one or more of these elements must be removed or disrupted.

- **Heat:** The fire needs to be heated to a certain temperature to sustain combustion.
- **Fuel:** Fires require a combustible material (wood, paper, gas, etc.) to burn.
- **Oxygen:** Fires need oxygen from the air to continue burning.

By removing or disrupting one or more of these elements, you can control or extinguish a fire.

2. Basic Fire Suppression Methods

There are several basic methods that can be used to suppress fires, depending on the size and type of the fire. Understanding when and how to apply these methods is key to keeping yourself, your family, and your home safe.

A. The P.A.S.S. Method for Extinguishers

A fire extinguisher is one of the most accessible and effective tools for suppressing small fires in an emergency. If you have access to an extinguisher, using the P.A.S.S. method will help you apply it correctly.

- **P (Pull):** Pull the pin from the handle of the fire extinguisher.
- **A (Aim):** Aim the nozzle at the base of the fire, not the flames.
- **S (Squeeze):** Squeeze the handle to discharge the extinguisher.
- **S (Sweep):** Sweep the nozzle from side to side to cover the entire fire's base.

Using this technique allows you to direct the flow of the extinguishing agent effectively, which is crucial for putting out the fire.

B. Smothering Fires

For smaller fires, such as those caused by grease or certain types of fuel, one of the simplest techniques is to cut off the fire's oxygen supply. Smothering involves covering the fire with a non-flammable material to deprive it of oxygen.

Grease Fires: Never use water on a grease fire, as it can spread the fire and make it worse. Instead, cover the fire with a metal lid, a wet cloth, or use a fire extinguisher rated for Class K fires (grease fires).

Clothing or Small Fires: For small fires on clothing or a person, use a heavy blanket, coat, or any large piece of fabric to smother the flames.

C. Using Water for Fires

Water is effective for suppressing many types of fires, particularly Class A fires (wood, paper, fabric, etc.), as it cools the fire and removes heat. However, it's crucial not to use water in situations where the fire involves electrical equipment (Class C), grease (Class K), or flammable liquids (Class B). In these situations, water can cause the fire to spread or lead to dangerous reactions.

Best for: Combustibles like wood, paper, or fabric.

Not for: Electrical, oil, or grease fires.

D. Fire Blankets

Fire blankets are simple yet effective tools that can be used to smother small fires, especially those on a person or on flammable liquids. Made from fire-resistant materials (such as fiberglass), fire blankets can be draped over a person or a fire to cut off the oxygen supply, putting out the flames.

Best for: Small fires, especially on clothing, or in confined spaces like kitchens.

3. Fire Suppression Tools and Equipment

While basic methods are helpful for small fires, larger fires require more specialized tools and equipment. Knowing what equipment to use and how to apply it can make the difference between a contained fire and a catastrophic disaster.

A. Fire Extinguishers

Fire extinguishers are essential for dealing with small to moderate fires. They come in several classes, each designed to handle different types of fires. Understanding which type of fire extinguisher to use is key to preventing the spread of the fire:

- **Class A**: For ordinary combustibles like paper, wood, and fabric.
- **Class B**: For flammable liquids, gases, and oils.
- **Class C**: For electrical fires.
- **Class D**: For combustible metals (used primarily in industrial settings).
- **Class K**: For kitchen fires involving cooking oils and fats.

Ensure that fire extinguishers are easily accessible and that family members are trained on how to use them properly. Regularly inspect and maintain your extinguishers to ensure they are in good working order.

B. Fire Hose or Nozzles

In more severe fire scenarios, a fire hose can be used to apply water at high pressure to contain or extinguish a fire. Fire hoses are typically only available in larger buildings or industrial settings, but they can be highly effective when available.

Best for: Large fires, particularly those that have spread beyond the capabilities of a small fire extinguisher.

C. Fire Sprinkler Systems

Fire sprinkler systems are permanently installed systems that detect heat and automatically discharge water to suppress fires. These systems are common in commercial and residential buildings and can be a vital line of defense in a fire emergency.

Best for: Large buildings, homes, and warehouses.

D. Fire Suppression Systems for Kitchens

For homes and businesses, kitchen fires are one of the most common threats. Special fire suppression systems are designed to prevent grease fires from escalating. These systems typically include hoods and nozzles that dispense fire-retardant chemicals directly onto the cooking surface.

Best for: Commercial kitchens, homes with heavy cooking, and restaurants.

E. Fire Retardant Chemicals and Foams

In certain scenarios, such as wildfires or industrial accidents, fire retardants and foams may be used to suppress fires. These chemicals work by either cooling the fire, coating the fuel to prevent ignition, or preventing the fire from spreading.

Best for: Large, out-of-control fires in industrial settings or outdoor wildfires.

4. Fire Safety Measures During an Emergency

While suppression techniques are important, preventing fire-related emergencies in the first place is equally crucial. Implementing fire safety measures in advance can reduce the likelihood of fires and provide better control if one does occur.

A. Fire Prevention Tips

Regular Inspection: Conduct regular inspections of electrical wiring, gas lines, and appliances to detect any signs of wear or malfunction that could lead to a fire.

Keep Fire Extinguishers Accessible: Ensure that fire extinguishers are easily accessible in key areas of the home, especially in kitchens, near fireplaces, and in garages.

Avoid Flammable Materials: Keep flammable materials (such as paper, cloth, and cleaning products) away from heat sources.

B. Creating Fire Breaks

In wildfire-prone areas, creating fire breaks is an effective strategy for preventing fires from spreading. Fire breaks are cleared areas devoid of vegetation, which can stop or slow the spread of fire.

How to Create Fire Breaks: Clear leaves, branches, and other debris from around your property and keep a defensible space between your home and any potential sources of fire.

C. Emergency Fire Plans

In the event of a fire, having a well-prepared emergency fire plan is essential. Every family member should know the evacuation routes, the location of fire extinguishers, and the safety procedures for different fire scenarios.

Establish Escape Routes: Identify at least two escape routes from each room, and ensure all windows and doors are easily operable in case of a quick exit.

Practice Drills: Regularly practice fire drills to ensure everyone knows how to react quickly and safely during a fire emergency.

In an emergency situation, your ability to suppress and manage fires can mean the difference between life and death. By understanding fire behavior, having the proper tools, and following safe suppression techniques, you can mitigate the risk and impact of fires. It's essential to prepare in advance, maintain fire safety practices, and regularly inspect equipment to ensure it is ready to use in an emergency. Whether it's through simple methods like smothering a small fire or employing advanced systems like sprinklers and fire suppression foams, these fire suppression techniques are indispensable tools for ensuring your survival during a crisis.

13. FINANCIAL AND LEGAL READINESS

Securing Critical Documents in a Crisis

In any crisis situation, having access to critical documents is vital for ensuring safety, securing resources, and navigating legal, financial, and personal challenges. Whether it's a natural disaster, civil unrest, or a personal emergency, the ability to quickly access important paperwork can determine the outcome of your survival and recovery. The proper handling, storage, and security of these documents can make a significant difference in your ability to prove identity, access funds, claim insurance, and rebuild your life.

1. Understanding Which Documents Are Critical

Before diving into how to secure documents, it is essential to understand which ones are most critical in a crisis. The specific needs will vary depending on the situation, but in general, the following documents are vital for handling emergencies:

- **Identification Documents:** These include birth certificates, Social Security cards, passports, driver's licenses, and state-issued IDs. These documents prove your identity and citizenship, which is crucial for accessing government assistance, medical care, and financial resources.
- **Insurance Policies:** Homeowners or renters insurance, health insurance, life insurance, car insurance, and flood or fire insurance policies are necessary for filing claims in the event of damage or loss.
- **Legal Documents:** Wills, powers of attorney, medical directives, marriage and divorce certificates, and custody agreements are critical for protecting your legal rights, ensuring healthcare decisions are respected, and securing custody of children or pets.
- **Financial Documents:** Bank account information, credit card statements, retirement account details, and tax records are essential for managing finances, accessing emergency funds, and proving financial eligibility for aid.
- **Medical Records:** This includes health records, prescriptions, vaccination records, and medical history, which can be essential for medical treatment during a crisis or when accessing healthcare.
- **Property Documents:** Titles and deeds for homes, vehicles, and other valuable property help prove ownership and may be needed for insurance claims or during recovery.
- **Emergency Plans and Contacts:** Emergency contact lists, evacuation plans, and any documents related to survival plans can be invaluable in a disaster situation.

2. Challenges of Document Security During a Crisis

In times of crisis, documents are often at risk due to a variety of factors:

Physical Damage: Fires, floods, and earthquakes can destroy paper documents. Even without major disasters, documents kept in attics, basements, or other vulnerable areas may deteriorate over time.

Theft or Loss: In times of chaos, there is an increased risk of theft or losing track of documents when you are forced to evacuate quickly.

Digital Accessibility: If your documents are stored electronically, you may not be able to access them if the power goes out or your digital storage systems fail. Cyber-attacks may also compromise sensitive data.

Displacement: If you are forced to evacuate or relocate due to a crisis, having to leave behind critical documents or not being able to access them quickly can severely hamper your recovery process.

3. Methods for Securing Documents

Ensuring that critical documents are safe, accessible, and intact in a crisis requires careful planning, organization, and use of various tools. Here are some strategies for securing both physical and digital copies of important documents:

A. Create Physical and Digital Copies

While keeping original documents safe is important, it's equally crucial to have copies that can be accessed easily if necessary.

- **Physical Copies:** Keep printed copies of your critical documents in a safe, fireproof, waterproof, and tamper-proof container. This container could be:
- **Fireproof Safe:** Choose a high-quality fireproof safe that is rated for both fire and water resistance. It should be large enough to store important papers such as birth certificates, social security cards, insurance policies, and property deeds.
- **Waterproof Bags:** For less valuable documents or for documents that need to be portable, a high-quality waterproof document bag can protect papers from water damage.
- **Lockbox:** A lockbox can be used for safe storage and quick access, particularly for essential papers like IDs and insurance policies.
- **Digital Copies:** Digitizing critical documents and storing them securely is increasingly essential in today's digital age. Options include:
- **Cloud Storage:** Use a secure, encrypted cloud storage service (such as Google Drive, Dropbox, or iCloud) to store digital copies of your documents. Make sure to use strong passwords and enable two-factor authentication (2FA) to protect your data.
- **Encrypted USB Drives:** For highly sensitive documents, such as financial information and health records, you can store digital copies on an encrypted USB drive. This ensures that, even if the drive is lost or stolen, your documents will remain secure.
- **Offline Digital Copies:** In the event of an internet outage, it's helpful to have offline copies of your documents saved on external hard drives or encrypted SD cards.

B. Use Document Scanning Apps

There are several smartphone apps that allow you to scan and store documents securely. Apps like Evernote, CamScanner, and Adobe Scan allow you to scan and store documents digitally. Make sure that the app you choose encrypts the data and backs up to the cloud.

Benefit: Having a portable and easy way to scan and store documents directly from your phone means you can quickly back up documents while on the go. Apps that use Optical Character Recognition (OCR) can also allow you to search and index the scanned documents for easier retrieval.

C. Backup Storage Locations

Never rely on a single storage method or location for your critical documents. It's important to have backups stored in different places to minimize the risk of losing everything in the event of theft, damage, or natural disaster.

- **Safe Deposit Box:** A safe deposit box in a bank can serve as a secure location for storing important documents, including titles, deeds, and insurance policies.
- **Trusted Family or Friend:** Consider giving copies of your documents to a trusted family member, friend, or lawyer. This ensures that you have access to them even if your primary location is compromised.
- **Diversified Cloud Services:** Using multiple cloud storage services ensures that even if one service is inaccessible due to an outage or issue, you can still access your documents from another.

4. Strategies for Organizing Critical Documents

The way you organize your critical documents can make a huge difference in your ability to find and access them during an emergency. Consider the following tips:

- **Categorize and Label:** Group your documents by category, such as Identification, Legal Documents, Insurance, and Financial. Use folders or labeled binders to separate them.
- **Create an Index:** Keep an index or master list that outlines where each document is stored, whether physically or digitally. This can include details about the storage location and passwords for encrypted files.
- **Use a Document Management System:** If you have a large number of important documents, a digital document management system (DMS) can help you organize, index, and retrieve them quickly.
- **Emergency Document Folder:** Keep an emergency folder or binder containing the most critical documents you may need in an urgent situation. This should include identification, insurance policies, medical records, and emergency contacts.

5. Additional Considerations

- **Password and Access Security:** If storing digital documents, ensure that all accounts and devices are secured with strong passwords, two-factor authentication, and encryption. For physical documents, consider limiting access to only trusted individuals.
- **Regular Updates:** Periodically review your documents to ensure they are up-to-date. Make sure that contact information, legal documents, and insurance policies are current and reflect any changes.
- **Disaster Drills:** Conduct practice runs of your emergency plan, including retrieving and accessing critical documents. This helps ensure that everyone in your household knows where to find the documents in an emergency and how to use them.

Securing critical documents is an essential part of disaster preparedness. The right combination of physical and digital storage, organization, and backup strategies ensures that you can quickly access and protect your vital information during a crisis. Properly secured documents help facilitate identification, access to financial resources, insurance claims, legal rights, and healthcare. By preparing and organizing your critical documents in advance, you improve your chances of navigating through emergencies successfully and ensure a smoother recovery in the aftermath.

Creating a Financial Emergency Fund

A financial emergency fund is a crucial pillar in any comprehensive crisis management strategy. Whether you're facing an unexpected job loss, a medical emergency, a natural disaster, or even a global pandemic, having a financial cushion to fall back on can make all the difference in your ability to weather the storm and maintain stability. The key to successfully managing personal finances during a crisis is preparation, and an emergency fund provides the essential financial safety net for unpredictable situations.

1. What Is an Emergency Fund?

An emergency fund is a savings account specifically set aside to cover unexpected expenses or emergencies, such as medical bills, home repairs, or the loss of income. It is typically liquid (easily accessible), so you can use the funds immediately when you need them. The purpose of the fund is to help you manage unforeseen financial burdens without relying on credit cards, loans, or other forms of debt, which can compound financial stress and hardship.

The key difference between an emergency fund and other types of savings is its focus on unforeseen, urgent expenses. The fund is not intended for planned or regular expenses (like vacations or home improvements) but for crises that threaten your financial security.

2. Why Do You Need an Emergency Fund?

An emergency fund is critical for maintaining financial stability during times of uncertainty. Without one, individuals often turn to credit cards, high-interest loans, or borrowing from family and friends, all of which can lead to long-term financial strain. Here are several reasons why an emergency fund is essential:

A. Protection Against Unexpected Expenses

Emergencies often arise without warning. You might be faced with a large medical bill, car repair, urgent home repair, or a sudden layoff from work. Without an emergency fund, these unexpected expenses can force you to dip into savings that were intended for other financial goals, or worse, accumulate high-interest debt.

B. Peace of Mind and Reduced Stress

Knowing that you have money set aside for emergencies can significantly reduce anxiety and stress. During times of crisis, the last thing you want is to add financial worry to your list of concerns. A financial safety net allows you to face emergencies with a clearer mind and a level of security.

C. Preventing Debt Accumulation

An emergency fund can prevent the need to rely on credit cards or loans, which often come with high-interest rates and fees. By having immediate access to cash, you can manage expenses without taking on additional debt. This can help you avoid a debt spiral, which may become difficult to escape from, especially in times of economic downturn.

D. Financial Flexibility and Independence

Having an emergency fund gives you more control over your financial decisions. For example, if you're faced with a job loss, an emergency fund can buy you time to find a new job or pursue other income-generating activities without the pressure to take the first offer that comes your way. It also allows you to make thoughtful decisions during uncertain times, rather than reacting out of desperation.

3. How Much Should You Save in Your Emergency Fund?

The amount to save for an emergency fund depends on several factors, including your monthly expenses, lifestyle, and the type of emergency you're preparing for. The general guideline is to have enough to cover three to six months' worth of living expenses. However, this amount can vary depending on your circumstances:

A. Standard Rule: Three to Six Months' Worth of Living Expenses

For most people, a good rule of thumb is to save enough to cover 3–6 months of living expenses. This provides a buffer for most financial emergencies, such as medical bills, home repairs, or temporary job loss. Here's how to calculate it:

- **Track Your Monthly Expenses**: Make a detailed list of all your monthly expenses, including rent or mortgage payments, utilities, groceries, transportation, insurance, and any other essential costs. This total will give you a baseline for how much you need to cover in case of an emergency.
- **Multiply by Three to Six**: Once you have your monthly expense total, multiply it by three to six months. This range provides a safety net for typical emergencies. For example, if your monthly expenses are $3,000, your emergency fund should ideally be between $9,000 and $18,000.

B. Special Circumstances

- **Freelancers and Self-Employed Individuals**: If you are self-employed or have an irregular income, it is recommended to save more, perhaps up to 12 months of living expenses. The unpredictable nature of freelance work or self-employment can make it harder to predict when and if the next paycheck will arrive.
- **Families with Dependents**: Families with children, elderly dependents, or others who rely on their income may need to save more, as the potential expenses during an emergency can be higher.
- **People with Health Issues**: If you or a family member has a chronic health condition or is prone to medical emergencies, it's prudent to save more, as medical bills can be costly.

4. Where Should You Keep Your Emergency Fund?

The safety of your emergency fund is just as important as how much you save. The fund should be easily accessible but not too easy to access, so it's not spent on impulse purchases. Consider the following options:

A. High-Yield Savings Accounts

High-yield savings accounts offer higher interest rates than standard savings accounts, helping your emergency fund grow over time while keeping it liquid. Many online banks offer competitive rates, and these accounts provide immediate access to your funds through a debit card or electronic transfer. Be mindful of any withdrawal restrictions or fees, but for most people, this is an excellent option for emergency savings.

B. Money Market Accounts

Money market accounts are another option for an emergency fund, offering slightly higher interest rates than regular savings accounts, along with easy access to your funds. While they may require a higher minimum balance than standard savings accounts, they offer liquidity while ensuring your funds are relatively safe.

C. Certificates of Deposit (CDs)

While not as liquid as savings accounts, certificates of deposit (CDs) can offer higher interest rates in exchange for locking your money up for a set period of time. You should only use CDs for emergency funds if you're comfortable with the possibility of having to wait until the CD matures, as early withdrawals typically incur penalties.

D. Cash at Home (Very Limited)

In some cases, having a small amount of cash available at home for emergencies can be useful, particularly for situations where electronic payments aren't possible. However, this should only be a small portion of your emergency fund, as cash stored at home isn't earning interest and can be easily lost or stolen.

5. How to Build Your Emergency Fund

Building an emergency fund may feel overwhelming, but by following a few steps, you can steadily grow your savings without putting a strain on your finances. Here are some strategies:

A. Set a Target and Break It Down

Start by setting a target for your emergency fund based on your monthly expenses. Break it down into smaller, more manageable goals. For example, if your goal is $12,000, aim to save $1,000 per month for a year. Having smaller, incremental targets can make the process feel less daunting.

B. Automate Your Savings

One of the easiest ways to build your emergency fund is to automate your savings. Set up an automatic transfer from your checking account to your emergency fund account each month. Treat this like any other monthly bill, and make sure to prioritize it to ensure steady growth.

C. Cut Unnecessary Expenses

Evaluate your monthly expenses to see if there are any non-essential purchases you can cut back on temporarily. Redirect those savings into your emergency fund. Even small changes, like eliminating subscription services you don't need or cooking meals at home instead of dining out, can add up over time.

D. Use Windfalls Wisely

Whenever you receive unexpected money, such as a tax refund, bonus, or gift, consider using a portion of it to boost your emergency fund. While it's tempting to spend windfalls on immediate gratification, putting that money toward your emergency fund can help you reach your goal faster.

E. Monitor and Adjust

Periodically assess your progress toward your goal and adjust as necessary. If you encounter unexpected expenses or income fluctuations, consider recalibrating your savings plan to stay on track.

Creating a financial emergency fund is one of the most important steps you can take to safeguard your financial future and ensure you're prepared for life's unpredictable events. By establishing a fund that can cover at least three to six months of living expenses, you'll have the security and peace of mind needed to face unexpected emergencies without falling into debt. Whether it's for medical emergencies, natural disasters, or job loss, having a well-funded emergency savings account is essential to maintaining your financial stability in times of crisis.

Preparing for Economic Disruptions

Economic disruptions—such as recessions, inflation spikes, or financial crises—can significantly impact personal and business finances. Proactive preparation is essential to mitigate these effects and maintain financial stability. This comprehensive guide outlines strategies to prepare for economic disruptions, with a particular focus on the current economic climate in Nigeria.

1. Understanding Economic Disruptions

Economic disruptions refer to significant, often sudden, changes in the economic environment that can affect individuals, businesses, and governments. These disruptions can manifest as:

Recessions: Periods of economic decline characterized by reduced economic activity, increased unemployment, and decreased consumer spending.

Inflation: A sustained increase in the general price level of goods and services, leading to a decrease in purchasing power.

Financial Crises: Events that cause a severe disruption in the financial markets, such as banking crises or stock market crashes.

Recognizing the signs of potential economic disruptions allows for timely preparation and response.

2. Assessing Personal Financial Health

Before implementing specific strategies, evaluate your current financial situation:

- **Income Stability:** Assess the reliability and diversity of your income sources.
- **Debt Levels:** Review outstanding debts and interest rates to identify areas for reduction.
- **Savings and Investments:** Examine the liquidity and risk levels of your savings and investment portfolios.

Understanding these aspects provides a foundation for targeted preparation efforts.

3. Building an Emergency Fund

An emergency fund is a critical component of financial preparedness:

- **Target Amount:** Aim to save three to six months' worth of living expenses. This buffer can cover unexpected costs during economic downturns.
- **Accessibility:** Keep the fund in a liquid, low-risk account, such as a high-yield savings account, to ensure quick access when needed.
- **Regular Contributions:** Set up automatic transfers to consistently build the fund over time.

In Nigeria for example, where inflation has been rising, with the annual rate reaching 34.60% in November 2024, maintaining an emergency fund is particularly crucial to preserve purchasing power.

4. Diversifying Income Streams

Relying on a single income source can be risky during economic disruptions:

- **Side Businesses:** Explore opportunities for additional income streams, such as freelancing or consulting.
- **Investments:** Consider investments in assets that can generate passive income, such as real estate or dividend-paying stocks.
- **Skill Development:** Enhance skills that are in demand to increase employability and potential earnings.

Diversification reduces the impact of a downturn in any single income source.

5. Reducing and Managing Debt

High debt levels can exacerbate financial strain during economic disruptions:

- **Prioritize High-Interest Debt:** Focus on paying off debts with the highest interest rates first to reduce overall financial burden.
- **Debt Consolidation:** Consider consolidating debts to secure lower interest rates and simplify payments.
- **Avoid New Debt:** Refrain from taking on new debt unless absolutely necessary.

6. Reviewing and Adjusting Investments

Economic disruptions can affect various investment vehicles:

- **Portfolio Diversification:** Ensure your investment portfolio is diversified across different asset classes to spread risk.
- **Risk Assessment:** Regularly assess the risk levels of your investments and adjust them according to your risk tolerance and market conditions.
- **Professional Advice:** Consult with financial advisors to make informed decisions during volatile periods.

In the face of rising economic uncertainty, experts recommend reviewing fixed-income investments for rate risk, as rising interest rates can decrease bond values.

7. Creating a Budget and Monitoring Expenses

A well-structured budget helps manage finances effectively:

- **Track Spending:** Monitor daily expenses to identify and eliminate unnecessary costs.
- **Essential vs. Non-Essential:** Differentiate between essential and non-essential expenses to prioritize spending.
- **Adjust for Inflation:** Account for rising costs in your budget to maintain purchasing power.

8. Enhancing Financial Literacy

Improved financial knowledge enables better decision-making:

- **Education:** Engage in financial literacy programs and workshops.
- **Resources:** Utilize reputable financial resources, such as books, online courses, and seminars.
- **Stay Informed:** Keep abreast of economic trends and news to anticipate potential disruptions.

Understanding the root causes of economic crises, such as those affecting Nigeria, can inform better financial decisions.

9. Developing a Contingency Plan

Having a plan in place prepares you for unforeseen events:

- **Scenario Planning:** Consider various economic scenarios and develop response strategies.
- **Resource Allocation:** Identify essential resources and ensure their availability during disruptions.
- **Communication Plan:** Establish a communication plan with family members or business partners to coordinate during crises.

10. Staying Informed and Flexible

Economic conditions are dynamic:

- **Continuous Monitoring:** Regularly review economic indicators and news to stay informed.
- **Adaptability:** Be prepared to adjust your strategies in response to changing economic conditions.
- **Community Engagement:** Participate in community discussions and support networks to share information and resources.

Preparing for economic disruptions requires proactive planning, financial discipline, and continuous education. By building an emergency fund, diversifying income sources, managing debt, and staying informed, individuals can enhance their resilience against economic challenges. In Nigeria, where recent economic reforms have led to increased living costs and inflation, these strategies are particularly pertinent.

14. LIVING THROUGH LONG-TERM CRISES

Sustaining Morale and Motivation During Economic Disruptions

Economic disruptions, such as recessions, financial crises, or periods of high inflation, can significantly impact individual and organizational morale and motivation. In Nigeria, where inflation reached 34.60% in November 2024, maintaining morale is particularly crucial. citereuters.com

1. Open and Transparent Communication

Clear communication fosters trust and reduces uncertainty. Regular updates about the organization's status, challenges, and future plans help employees feel informed and valued. Encouraging feedback and addressing concerns promptly can alleviate anxiety and build a supportive environment. citetheemployeeapp.com

2. Recognize and Appreciate Efforts

Acknowledging individual and team contributions boosts morale. Simple gestures like verbal praise, personalized notes, or public recognition can make employees feel valued, especially during challenging times. citetheemployeeapp.com

3. Provide Emotional Support

Offering access to mental health resources and creating a supportive atmosphere helps employees manage stress and maintain motivation. Encouraging open discussions about well-being and providing support systems can enhance resilience. citepbahealth.com

4. Foster a Positive Work Environment

Cultivating a culture of respect, collaboration, and positivity can counteract the negative effects of economic stress. Organizing team-building activities, even virtually, can strengthen bonds and improve morale. citeorderry.com

5. Offer Flexibility and Support

Providing flexible work arrangements and understanding personal challenges during economic disruptions can reduce stress and increase job satisfaction. Demonstrating empathy and support for employees' personal situations fosters loyalty and motivation. citepbahealth.com

6. Encourage Professional Development

Investing in employees' growth through training and development opportunities shows commitment to their future, enhancing motivation and engagement. Providing resources for skill development can empower employees and improve performance. citethepolyglotgroup.com

7. Celebrate Small Wins

Recognizing and celebrating small achievements can boost morale and motivation. Acknowledging milestones, even minor ones, helps maintain a sense of progress and accomplishment. citeorderry.com

8. Maintain Work-Life Balance

Encouraging employees to balance work with personal time prevents burnout and maintains motivation. Promoting a healthy work-life balance is essential for long-term productivity and well-being. citepbahealth.com

9. Provide Clear Direction and Purpose

Aligning individual roles with the organization's mission and goals helps employees see the value of their work, enhancing motivation. Clear expectations and a sense of purpose can drive engagement and performance. citethepolyglotgroup.com

10. Lead by Example

Leadership plays a crucial role in setting the tone for morale. Demonstrating resilience, positivity, and commitment can inspire employees to adopt similar attitudes. Effective leadership fosters a motivated and engaged workforce. citethepolyglotgroup.com

By implementing these strategies, organizations can sustain morale and motivation during economic disruptions, leading to improved performance and resilience.

Adjusting to a New Normal During Economic Disruptions

The global economy is experiencing continuous shifts, driven by a variety of factors such as pandemics, technological advancements, geopolitical tensions, and environmental changes. As a result, both individuals and organizations are increasingly faced with the challenge of adjusting to what is now being called a "new normal." This adjustment process is often especially critical during economic disruptions, as it requires resilience, adaptability, and a proactive approach to navigating uncertainty.

1. Understanding the New Normal

The "new normal" refers to a shift in societal behavior, economic conditions, and business operations that emerges after a major disruptive event. In the wake of economic disruptions like recessions, inflation spikes, or global crises, this shift can feel particularly disorienting. Individuals may find themselves facing job insecurity, reduced income, or higher living costs, while businesses may experience changes in demand, workforce constraints, and shifting customer expectations.

According to a 2024 report from the International Monetary Fund (IMF), global economies are facing a prolonged period of slow recovery following the COVID-19 pandemic and the economic ripple effects caused by supply chain disruptions, inflation, and geopolitical instability. The new normal encompasses not only economic changes but also alterations in social norms and individual behavior. People are prioritizing sustainability, local production, and long-term stability in response to these challenges.

2. Economic Shifts and Their Impacts

The economic landscape has fundamentally changed, with some of the key trends including:

Increased Inflation: Rising prices for essential goods, including food, fuel, and housing, have stretched household budgets. A survey conducted by the U.S. Federal Reserve in 2024 found that inflation has led to a 30% increase in basic living expenses for most American families over the past two years.

Job Market Transformation: Many industries are embracing automation and remote work, leading to job displacement in certain sectors but also creating new opportunities in others. According to a McKinsey & Company report, industries such as healthcare, technology, and renewable energy are thriving, while traditional manufacturing sectors are facing difficulties.

Supply Chain Instability: Supply chain disruptions, which began during the COVID-19 pandemic, have continued, affecting everything from food distribution to electronics. According to a 2024 report from the World Economic Forum, companies are now investing heavily in resilient and localized supply chains to mitigate these issues in the future.

3. Adjusting at the Personal Level

The personal adjustment to a new normal often revolves around managing financial stress, adapting to new work patterns, and rethinking consumption habits. Here are several strategies individuals can use to navigate economic disruptions:

a) Financial Resilience and Flexibility

In times of economic disruption, individuals may need to rethink their financial strategies. Creating a financial buffer by saving a larger emergency fund or diversifying income sources is vital. According to a 2023 study by the Pew Research Center, nearly 50% of U.S. households reported a significant change in their spending habits due to rising costs. Many people are shifting to budgeting for essentials, reducing discretionary spending, and embracing frugal habits like couponing or buying in bulk.

Additionally, investing in skills development or learning new trades can provide more flexibility and security. For instance, platforms like Coursera and Udemy offer affordable courses on topics ranging from digital marketing to project management, which can help individuals shift careers or boost their earning potential in uncertain times.

b) Adapting to Remote Work

The rise of remote work is one of the most prominent shifts brought about by economic disruptions and the COVID-19 pandemic. In 2024, a survey by Gallup showed that 33% of the U.S. workforce is working remotely at least part-time, a trend that is expected to continue as many employers find the remote work model cost-effective and desirable for their employees' work-life balance.

Adjusting to this new work environment requires adopting new communication tools (e.g., Zoom, Slack) and rethinking work-life boundaries. Establishing a dedicated workspace at home, setting clear work hours, and managing distractions are all important steps in staying productive and maintaining mental well-being.

c) Emphasizing Mental Health and Well-Being

The stresses brought about by economic disruptions can have profound effects on mental health. Anxiety, depression, and burnout rates have risen significantly during these periods of upheaval. In 2024, the American Psychological Association reported that nearly 60% of adults in the U.S. are experiencing heightened stress due to economic uncertainties.

To manage stress, individuals are increasingly turning to mental health support such as therapy, meditation, and exercise. Online therapy platforms like BetterHelp and Talkspace are providing affordable and convenient access to professional counseling. Additionally, mindfulness practices, such as yoga or journaling, have proven to be effective in managing stress and improving mental health.

d) Revising Consumption Habits

As inflation drives up costs, people are adjusting their consumption habits to reflect a more sustainable and value-driven approach. Many are opting for second-hand goods, repairing instead of replacing items, and focusing on purchasing locally-sourced products to support their communities. Consumer behavior studies in 2024 indicate a sharp rise in demand for sustainable products and services as individuals become more conscious of their economic impact and environmental footprint.

4. Adapting at the Organizational Level

Organizations also need to adjust to the new normal, rethinking how they operate, communicate, and engage with their employees and customers.

a) Adopting Flexible Business Models

Organizations are increasingly moving towards flexible business models to thrive in a disrupted economic environment. For instance, businesses are implementing hybrid work arrangements, which not only provide flexibility but also help them tap into a broader talent pool. According to Deloitte's 2024 Global Workforce Survey, 75% of executives are considering or have already adopted hybrid models to maintain operational efficiency.

Additionally, businesses are diversifying their revenue streams to mitigate the risk of relying on one segment. For instance, restaurants may begin offering meal kits or delivery services alongside traditional dining, while retail stores may integrate online shopping with in-store experiences to accommodate customer preferences.

b) Sustainability and Resilience

Businesses are increasingly investing in sustainability and resilience practices, not only to align with consumer preferences but also to future-proof their operations. According to a report from the World Economic Forum, 60% of companies surveyed in 2024 said that climate change risks and disruptions were the primary factors driving sustainability efforts. This includes investing in renewable energy, reducing waste, and adopting eco-friendly materials.

c) Nurturing Employee Well-Being

In the wake of economic disruptions, companies are focusing on employee well-being and mental health. Flexible work arrangements, paid time off, and employee assistance programs are becoming essential components of the workplace culture. A 2024 report from Gallup highlighted that companies with strong employee well-being programs saw increased productivity and employee retention.

5. Building a More Resilient Future

Adjusting to the new normal during economic disruptions requires long-term thinking, flexibility, and a focus on building resilience. Both individuals and organizations must embrace change, invest in personal and professional development, and prioritize mental health and sustainability. By making strategic adjustments now, individuals and organizations can better navigate the uncertainties of the future and emerge stronger, more adaptable, and better prepared for any challenges that may arise.

In the coming years, those who have proactively adapted to the new normal will be better equipped to thrive amidst economic disruptions, while those who resist change may find themselves struggling to keep up. Building resilience is no longer just a strategy—it is a necessity.

Teaching Survival Skills to Children

Survival skills are essential tools for self-reliance, and while adults often take the lead in preparing for emergencies, it's increasingly important to include children in the process. Teaching survival skills to children not only empowers them with the knowledge and confidence to handle emergencies but also fosters a sense of responsibility, independence, and respect for nature. Whether you're preparing for natural disasters, economic disruptions, or other unforeseen events, children who are equipped with basic survival skills will be better prepared to manage unexpected challenges.

This guide outlines essential survival skills for children and provides practical tips for teaching these skills in a fun, engaging, and age-appropriate manner. Understanding how to integrate survival lessons into daily life will give your child a stronger foundation for thriving in adverse situations.

1. Why Teaching Survival Skills to Children is Important

In an unpredictable world, children need more than just book knowledge; they need practical, hands-on experience. Survival skills provide children with the ability to think critically, solve problems, and stay calm in stressful situations. Whether it's knowing how to find food and water, navigate without GPS, or build a shelter, these skills offer a sense of security and independence.

Moreover, introducing these skills early can cultivate positive habits and reinforce important life lessons:

Self-Reliance: Children learn the importance of taking care of themselves and others.

Teamwork: Many survival situations involve collaboration, teaching children how to work together in challenging circumstances.

Critical Thinking: Survival situations often require quick thinking, resourcefulness, and creativity, fostering problem-solving abilities in children.

Respect for Nature: Children who understand the natural world are more likely to value and protect the environment.

2. Basic Survival Skills for Children

Teaching survival skills to children should be age-appropriate, starting with simple concepts and gradually building to more complex tasks. Here are some foundational survival skills to teach children, tailored to various age groups.

a) Basic Safety Rules

For young children, the first step in survival training should be understanding basic safety rules. This includes:

Fire Safety: Teaching children how to safely use matches and lighters, as well as what to do in case of a fire. Fire drills at home, showing children how to stop, drop, and roll, and understanding escape routes can help children prepare in emergencies.

Stranger Danger: Familiarizing children with the importance of staying safe in unfamiliar situations and what to do if they get separated from family.

First Aid Basics: Basic first aid skills, like knowing how to clean and bandage a cut, can be taught at a young age through role-playing and first aid kits designed for kids.

b) Water Safety and Survival

Water is one of the most essential survival resources. Teaching children how to manage water supplies and stay safe around water bodies is critical:

How to Find Water: Explain the concept of water purification (e.g., using a filter or boiling water) in a way children can understand, making sure they know that drinking untreated water is dangerous.

Swimming and Floating: If your child is old enough, teaching them how to swim or float in water can be life-saving. Enroll them in swimming classes or take lessons together to make learning fun.

c) Building a Shelter

For older children, understanding how to build a simple shelter could be vital in the event of a natural disaster or while camping in the wilderness:

Shelter Building: Depending on their age, you can teach children how to gather materials (leaves, branches, tarps) to construct a shelter. Start small by creating forts or small shelters in your backyard to practice.

Using a Tent: Children should also understand how to set up a tent properly. This is a hands-on skill that can be practiced during family camping trips.

d) Fire Building

Fire is a basic necessity for warmth, cooking, and signaling for help in emergency situations:

Fire Building Basics: Once children are old enough (usually 8-10 years), teach them how to start a fire using a fire starter, matches, or even a magnifying glass. Always supervise, emphasizing fire safety.

Fire-Starting Materials: Explain the different types of materials needed to start a fire: tinder, kindling, and fuel wood. Let children gather these materials to practice starting small, controlled fires with guidance.

e) Navigation and Orienteering

Older children can benefit from learning basic navigation skills, which are invaluable during outdoor adventures or emergencies:

Using a Compass and Map: Teach children how to read a map and use a compass to find directions. Have them practice navigating short distances in the neighborhood or local park.

Landmarks and Natural Indicators: Children can learn to use the sun, stars, and natural features (such as trees and rivers) for basic navigation.

3. Making Survival Training Fun and Engaging

Children learn best when they're engaged and having fun. To keep them motivated and interested, consider incorporating these activities into their survival training:

a) Survival Games

Treasure Hunts: Organize a scavenger hunt where children need to use clues to find essential survival items like a first aid kit, a flashlight, or a whistle.

Wilderness Challenges: Create mock survival scenarios where kids have to solve problems like finding food, building a shelter, or making a fire. This hands-on approach helps children understand the practical applications of survival skills.

Role-Playing: Role-playing different survival situations (e.g., becoming lost in the woods or handling an emergency at home) can help children practice staying calm under pressure and making decisions.

b) Survival Kits for Kids

Creating a child-sized survival kit can be a fun and educational activity. Let your child help assemble the kit, which could include:

- A whistle
- A flashlight or headlamp
- A small first-aid kit
- A multi-tool or small knife (for older children)
- A simple compass

- Emergency snacks (e.g., granola bars, trail mix)
- A poncho or emergency blanket
- A notepad and pen for taking notes or leaving messages

c) Incorporating Survival Skills into Everyday Activities

Survival skills don't need to be taught in isolation. Incorporating lessons into daily activities will make them feel more natural and useful:

Cooking: Teach your child how to prepare simple meals using ingredients you already have. This can be a valuable skill, especially in situations where you need to make the most out of limited supplies.

Gardening: Show children how to grow vegetables and herbs, emphasizing the importance of cultivating food in times of need. Kids can help plant and tend to a garden in your backyard, teaching them about food production and sustainability.

4. Instilling Confidence Through Experience

One of the most important aspects of teaching survival skills is to instill a sense of confidence and self-assurance in children. By exposing them to various scenarios and gradually increasing the complexity of the skills they learn, children become more capable and self-reliant.

Praise and Encouragement: When teaching survival skills, be sure to provide plenty of positive reinforcement. Celebrate small victories, such as successfully building a fire or finding their way with a compass.

Empathy and Emotional Resilience: It's also crucial to help children understand that sometimes, survival situations can be emotional as well as physical. Teaching children how to manage stress, practice patience, and work through fears is just as important as the technical skills they learn.

Teaching survival skills to children is an investment in their future resilience. It empowers them to think critically, make informed decisions, and respond to emergencies with confidence. Whether you're teaching basic fire-starting techniques or advanced navigation skills, remember that each lesson instills lifelong values of resourcefulness, self-reliance, and respect for the world around them.

By making these lessons fun, interactive, and age-appropriate, you can create an enriching learning experience that ensures the next generation is ready to handle whatever challenges life may throw their way.

15. CONCLUSION

The Importance of Ongoing Preparedness

In an era marked by rapid technological advancements, climate change, and geopolitical uncertainties, the importance of ongoing preparedness cannot be overstated. Preparedness involves proactive planning, training, and resource allocation to effectively respond to potential emergencies and disasters. By maintaining a state of readiness, individuals, communities, and organizations can mitigate risks, reduce the impact of unforeseen events, and expedite recovery processes.

1. Understanding Preparedness

Preparedness encompasses a range of activities designed to enhance the capacity to respond to emergencies. This includes developing emergency plans, conducting training exercises, establishing communication protocols, and ensuring the availability of necessary resources. The goal is to anticipate potential threats and implement measures that can prevent, mitigate, or effectively respond to them.

2. The Benefits of Ongoing Preparedness

Risk Mitigation: Regular preparedness activities help identify potential hazards and vulnerabilities, allowing for the implementation of measures to reduce or eliminate risks. For instance, updating safety procedures and securing equipment can prevent accidents and minimize damage during emergencies.

Enhanced Response Efficiency: Preparedness ensures that individuals and organizations are equipped with the knowledge and resources to respond swiftly and effectively to emergencies. This readiness can significantly reduce response times and improve outcomes during crises.

Resource Optimization: By planning and training in advance, resources can be allocated more efficiently during emergencies. This includes ensuring that equipment is maintained and that plans are reviewed and tested regularly, leading to cost savings and reduced downtime.

Community Resilience: Ongoing preparedness fosters a culture of resilience within communities, empowering individuals to act confidently and effectively during emergencies. This collective readiness enhances the overall capacity to withstand and recover from disasters.

3. Key Components of Ongoing Preparedness

Planning: Developing comprehensive emergency plans that outline roles, responsibilities, and procedures during crises. Strategic and operational planning establishes priorities, identifies expected levels of performance, and helps stakeholders understand their roles.

Training and Exercises: Regular training sessions and simulation exercises ensure that individuals and teams are familiar with emergency procedures and can perform effectively under pressure. This includes conducting drills and tabletop exercises to practice response strategies.

Resource Management: Ensuring the availability and maintenance of necessary resources, such as emergency kits, medical supplies, and communication tools, is crucial. Preparedness requires constant vigilance to ensure that equipment is maintained and that plans are reviewed and tested regularly.

Communication Systems: Establishing reliable communication channels to disseminate information during emergencies is vital. This includes setting up early warning systems and ensuring that all stakeholders are informed promptly.

4. Challenges in Maintaining Ongoing Preparedness

Complacency: Over time, the absence of disasters can lead to complacency, causing individuals and organizations to neglect preparedness activities. This "disaster dementia" can result in inadequate responses when emergencies occur.

Resource Constraints: Limited budgets and resources can hinder the implementation of comprehensive preparedness programs. Prioritizing preparedness within organizational budgets is essential to overcome this challenge.

Evolving Threats: The dynamic nature of threats, including climate change and emerging technologies, requires continuous reassessment and adaptation of preparedness plans. Regular updates and flexibility in planning are necessary to address new risks.

5. Global Perspectives on Preparedness

International organizations emphasize the significance of preparedness in disaster risk reduction. The United Nations Office for Disaster Risk Reduction (UNDRR) advocates for proactive measures to prevent, respond to, and recover from humanitarian emergencies. Their approach includes risk analysis, planning, and capacity building to enhance resilience.

6. The Role of Technology in Preparedness

Advancements in technology have transformed preparedness strategies. Tools such as geographic information systems (GIS), early warning systems, and mobile applications facilitate real-time data collection, analysis, and communication, thereby improving response times and coordination during emergencies.

7. The Preparedness Paradox

The preparedness paradox suggests that the more an individual or society prepares for a disaster, the less the harm if and when that event occurs. Because the harm was minimized, the people then wonder whether the preparation was necessary. This paradox can lead to complacency and underestimation of the importance of ongoing preparedness. citeturn0search11

Ongoing preparedness is a critical component of effective emergency management. By proactively planning, training, and resource allocation, individuals and organizations can enhance their resilience to a wide range of potential emergencies. In a world characterized by uncertainty, maintaining a state of readiness is not just beneficial but essential for safeguarding lives, property, and communities.

Building a Lifestyle of Readiness and Resilience

In today's dynamic world, cultivating a lifestyle centered on readiness and resilience is essential for effectively navigating life's challenges and uncertainties. This approach involves proactively developing the skills, habits, and mindsets that enable individuals to adapt, recover, and thrive amidst adversity.

1. Understanding Readiness and Resilience

- **Readiness:** The state of being prepared to respond effectively to potential challenges or emergencies. It encompasses having the necessary knowledge, skills, and resources to act promptly and efficiently when needed.
- **Resilience:** The capacity to recover quickly from difficulties; it's the ability to bounce back from setbacks, adapt to change, and keep going in the face of adversity. Resilience is not an innate trait but a set of skills that can be developed and strengthened over time.

2. Key Components of a Resilient Lifestyle

- **Physical Well-being:** Maintaining good health through regular exercise, balanced nutrition, and adequate rest is foundational. Physical fitness enhances energy levels, reduces stress, and improves overall mood, contributing to greater resilience.
- **Emotional Intelligence:** Developing the ability to recognize, understand, and manage one's emotions, as well as empathize with others, fosters emotional stability and effective interpersonal relationships.
- **Social Connections:** Building and nurturing a supportive network of family, friends, and community members provides a safety net during challenging times. Strong social ties are linked to better mental health and increased resilience.
- **Mental Agility:** Cultivating a growth mindset and the ability to adapt to new situations enhances problem-solving skills and the capacity to learn from experiences.
- **Purpose and Meaning:** Engaging in activities that align with personal values and passions fosters a sense of purpose, which is crucial for motivation and resilience.

3. Strategies to Build Readiness and Resilience

- **Set Clear Goals:** Establish both short-term and long-term objectives to provide direction and motivation. Clear goals help maintain focus and a sense of purpose.
- **Develop Problem-Solving Skills:** Engage in activities that challenge cognitive abilities, such as puzzles or strategic games, to enhance critical thinking and adaptability.
- **Practice Mindfulness and Stress Management:** Incorporate mindfulness practices, meditation, or deep-breathing exercises into daily routines to manage stress and maintain emotional balance.
- **Engage in Continuous Learning:** Pursue new skills and knowledge to build confidence and adaptability. Lifelong learning keeps the mind sharp and opens up new opportunities.
- **Maintain Financial Preparedness:** Develop and adhere to a budget, build an emergency fund, and plan for future financial needs to reduce stress during unexpected situations.

- **Foster Community Involvement:** Participate in local organizations or volunteer opportunities to strengthen social bonds and contribute to community resilience.

4. Overcoming Challenges in Building Resilience

- **Combatting Complacency:** Regularly reassess personal goals and preparedness plans to avoid stagnation. Engage in new experiences to keep the mind and body active.
- **Managing Negative Self-Talk:** Replace self-doubt with positive affirmations and realistic self-assessment to build confidence and resilience.
- **Balancing Work and Life:** Set boundaries to prevent burnout and ensure time for self-care and relaxation.

5. The Role of Organizations in Promoting Resilience

Organizations can play a pivotal role in fostering resilience among their members by:

- **Providing Training and Resources:** Offer programs that teach resilience skills, such as stress management workshops or leadership development courses.
- **Encouraging Open Communication:** Create an environment where individuals feel safe to express concerns and seek support.
- **Recognizing and Rewarding Resilience:** Acknowledge and celebrate instances where individuals or teams demonstrate resilience, reinforcing its value within the organization.

6. The Impact of Resilience on Longevity and Well-being

Recent studies have highlighted the significant role of resilience in enhancing longevity and overall well-being. For instance, a study published in BMJ Mental Health found that higher mental resilience scores were associated with a 53% reduction in the risk of death among older adults. This underscores the importance of cultivating resilience to improve quality of life and increase lifespan. citeturn0news11

Building a lifestyle of readiness and resilience is an ongoing process that requires intentional effort and commitment. By integrating the strategies outlined above into daily life, individuals can enhance their capacity to navigate challenges, adapt to change, and emerge stronger from adversity. Ultimately, fostering resilience not only improves personal well-being but also contributes to the strength and resilience of the broader community.

APPENDICES

Comprehensive Supply Checklists

A comprehensive supply checklist is an essential tool for any prepper, providing a clear and organized guide to ensure you are fully prepared for emergencies and survival scenarios. These checklists include essential categories of supplies that should be gathered, stored, and rotated regularly. Here's a breakdown of critical supplies to include in your emergency preparedness plan:

1. Food Supply Checklist

Food is the cornerstone of survival preparedness. You need to ensure that you have a well-rounded selection of non-perishable items that can sustain you for an extended period.

Short-Term Food Storage (1–3 months)

- Canned goods (vegetables, fruits, beans, soups)
- Rice, pasta, and grains
- Breakfast cereals and oatmeal
- Freeze-dried meals (ready-to-eat)
- Energy bars and granola bars
- Jerky (beef, turkey, or other proteins)
- Canned tuna, chicken, and other meats
- Powdered milk or evaporated milk
- Nut butters (peanut, almond)
- Ready-to-drink protein shakes

Long-Term Food Storage (3+ months)

- Freeze-dried fruits, vegetables, and meats
- Dehydrated soups and stews
- MREs (Meals Ready to Eat)
- Powdered eggs and powdered butter
- Long-term survival rations (e.g., Mylar-sealed meals)
- Bulk grains (wheat, oats, corn, quinoa)
- Salt, sugar, honey, and spices
- Long-term fruit and vegetable powders

2. Water Supply Checklist

Access to clean drinking water is critical for survival. Water purification methods and large storage systems are necessary to ensure you can remain hydrated during a crisis.

Water Storage

- Water barrels or large water containers (5-gallon, 50-gallon)
- Water purification tablets
- Water purification filters (e.g., LifeStraw, Sawyer Mini)
- Collapsible water containers (e.g., Camelbak, Platypus)
- Water purification pumps
- Water Filtration & Purification Tools
- Water filter systems (e.g., Berkey, MSR Guardian)
- Solar stills for water collection
- Water boiling equipment (camp stoves, portable kettles)
- Iodine and chlorine-based purification solutions

3. Medical Supplies Checklist

Having a well-stocked first-aid kit and access to necessary medical supplies is essential for dealing with injuries, illnesses, and emergencies.

Basic First-Aid Kit

- Adhesive bandages (various sizes)
- Sterile gauze pads and bandage rolls
- Antiseptic wipes and ointments (iodine, Neosporin)
- Pain relievers (ibuprofen, acetaminophen)
- Tweezers and scissors
- Medical tape
- Thermometer
- Cotton swabs and cotton balls
- Burn ointment and aloe vera gel
- Advanced Medical Kit
- Prescription medications (if applicable)
- Antibiotic ointments
- Antihistamines (for allergic reactions)
- Stitches or skin closure strips
- IV fluids or saline solution
- Trauma dressings and tourniquets
- Medical gloves and face masks
- Splints and slings
- Instant cold packs and heating pads
- EpiPen (for severe allergic reactions)
- Personal Hygiene and Sanitation Supplies
- Hand sanitizers and disinfectant wipes

Soap (bar or liquid)

Toothbrushes, toothpaste, and floss

Wet wipes and baby wipes

Toilet paper and portable toilet (if needed)

Feminine hygiene products

Towel and washcloth

4. Shelter & Warmth Checklist

Ensuring you have adequate shelter and warmth during emergencies is vital for your survival. Here are essential items for creating a safe, warm, and dry environment.

Emergency blankets (space blankets)

Sleeping bags (3-season or winter-rated)

Sleeping pads or air mattresses

Tents and tarps

Ground tarps for shelter setup

Camp stoves and fuel (propane, butane, or wood-burning)

Firestarter kits (matches, lighters, ferro rods)

Portable heaters (propane, catalytic, or wood stoves)

Extra clothing (thermal layers, waterproof jackets, hats, gloves, socks)

Rain gear and ponchos

5. Tools & Equipment Checklist

Tools are essential for building shelter, preparing food, and maintaining survival systems. The following tools are crucial for survival in a disaster or emergency situation.

Basic Tools

Multi-tool (e.g., Leatherman, Swiss Army Knife)

Axes and hatchets

Shovel (folding or tactical)

Crowbar and pry bar

Handsaw or folding saw

Duct tape and zip ties

Paracord (100 feet minimum)

Work gloves and protective gear

Climbing rope and carabiners

Lighting & Power Tools

Solar-powered flashlights and lanterns

LED headlamps

Solar power banks and chargers

Portable generators (gas, solar, or battery-powered)

Backup batteries (AA, AAA, 9V)

Solar panels or portable solar kits

6. Security & Self-Defense Checklist

In any emergency situation, security and personal safety are top priorities. These items will help you protect your home and loved ones.

Firearms and ammunition (if legal in your area)

Pepper spray or mace

Batons or tactical weapons

Personal alarm or whistle

Security cameras and motion detectors

Tactical flashlight with strobe function

Alarm system for windows and doors

Heavy-duty locks and deadbolts

Bulletproof vest (optional but recommended for certain crises)

7. Communication & Signaling Checklist

Effective communication can make all the difference in a survival situation. You need to stay connected with loved ones and emergency services while ensuring your safety.

Two-way radios (walkie-talkies) with extra batteries

Emergency radio (battery or hand-crank)

Signal mirrors for emergency signaling

Emergency whistle

Satellite phone (if needed)

Pre-written emergency contact list

Solar-powered or hand-crank phone chargers

8. Documentation & Important Papers Checklist

In a crisis, having access to key documents and identification is essential. These items should be securely stored but easily accessible.

Copies of personal identification (passport, driver's license, birth certificates)

Medical records (immunizations, prescriptions)

Insurance documents (home, health, car, life)

Financial records (bank account details, investment information)

Emergency contact information

Emergency plans (evacuation routes, meeting places)

Wills and legal documents

9. Financial Preparedness Checklist

Having access to financial resources during a crisis is crucial for purchasing necessities. Keep a cash reserve and financial assets accessible in various forms.

Cash in small denominations (especially if ATMs or digital payment systems are down)

Gold, silver, or precious metals (for barter or long-term wealth preservation)

Prepaid debit cards or emergency fund accounts

Essential credit card details for purchases when needed

10. Miscellaneous Items Checklist

There are numerous smaller, yet highly important, supplies that can be vital in specific situations.

Spare eyeglasses or contact lenses

Insect repellent and sunscreen

Maps of local area and surrounding regions

Notebooks and pens for record-keeping

Sewing kit for minor repairs

Heavy-duty trash bags (for waste management)

Spare keys for home, car, and safe

This comprehensive supply checklist ensures that you are prepared for nearly any emergency or crisis situation. Regularly reviewing and rotating your supplies will keep them fresh and ready for when you need them most. Don't forget to customize your list based on your specific needs, climate, and environment.

Resource Lists for Preppers

A well-rounded preparedness plan goes beyond just supplies. It also includes valuable resources that preppers can rely on for knowledge, tools, and systems that will help them survive and thrive during any emergency situation. Below is a comprehensive list of resources that preppers should consider integrating into their plans:

1. Survivalist and Preparedness Books

Books are invaluable resources for learning essential survival skills and strategies, from self-defense to sustainable living. A well-rounded prepper library should include guides on various survival topics.

Essential Survival and Preparedness Books

"The Encyclopedia of Country Living" by Carla Emery

A guide to homesteading, food preservation, and farming, with a focus on self-sufficiency.

"The Prepper's Blueprint" by Tess Pennington

A comprehensive guide to building a survival plan for any disaster scenario.

"Emergency War Surgery" by the U.S. Department of Defense

An essential resource for learning field medical techniques and trauma care.

"The SAS Survival Handbook" by John "Lofty" Wiseman

A classic on wilderness survival, navigation, and practical tips for living in extreme conditions.

"The Survival Medicine Handbook" by Joseph Alton, MD, and Amy Alton, ARNP

A comprehensive manual on first aid, medical emergencies, and field treatments.

"When All Hell Breaks Loose" by Cody Lundin

Focuses on psychological and practical skills for surviving through disasters.

2. Online Resources and Blogs

The digital world offers countless resources for prepping, from blogs and forums to educational videos and eBooks. Some of these are updated regularly with survival news, tips, and tactics.

Popular Prepper Websites

The Organic Prepper

A blog offering practical advice on food storage, alternative energy, and prepping for various emergencies.

Survival Blog

A resource rich with articles on survival skills, gear reviews, and prepping strategies.

Backdoor Survival

Covers prepping basics, gear, and how to live sustainably and self-sufficiently.

Modern Survival Blog

Focuses on preparedness strategies and self-sufficiency, offering valuable insights on bugging in, bugging out, and survival gear.

Prepper Website

A collection of articles, podcasts, and resources from experienced preppers about everything from food storage to firearm safety.

3. Podcasts and YouTube Channels

Podcasts and video channels are excellent resources for learning new skills, hearing interviews with experts, and staying informed on the latest trends in preparedness.

Prepper Podcasts

The Survivalist Prepper Podcast

Offers a comprehensive range of topics including survival skills, bug-out strategies, and prepping psychology.

The Prepping Academy Podcast

Covers prepping basics, reviews, and real-life experiences of preppers.

The Practical Preppers Podcast

Focuses on off-grid living, emergency preparedness, and sustainable living solutions.

The Preparedness Podcast

Discusses disaster scenarios and interviews with experts in different survival fields.

YouTube Channels

Canadian Prepper

Covers gear reviews, prepping advice, and survival strategies for all levels of preparedness.

Prepper Princess

Focuses on prepping on a budget, offering tips for storing food, supplies, and creating emergency plans.

The Urban Prepper

A great resource for city dwellers looking to prep in urban environments, including bug-out bag essentials and DIY projects.

SouthernPrepper1

A channel that provides prepping tips for all skill levels, as well as gear reviews and practical advice for surviving different types of crises.

4. Survival Gear Suppliers and Manufacturers

Having access to reliable suppliers for quality gear is essential for ensuring your preparedness plan is robust. These resources offer high-quality gear and products for various survival needs.

Key Gear Suppliers

REI

Offers a wide range of outdoor and survival gear, including tents, stoves, sleeping bags, and backpacks.

The Ready Store

A specialized prepper store offering long-term food storage, survival kits, and emergency gear.

Survival Frog

Provides prepping gear, survival kits, and emergency supplies.

Emergency Essentials

Specializes in food storage, water filtration, and preparedness supplies for emergencies.

Backcountry

Offers outdoor gear, survival equipment, and camping supplies ideal for prepping and bug-out scenarios.

5. Emergency Communications Tools

Communication during an emergency is vital for coordinating with loved ones, getting updates on the situation, and receiving help if needed. Investing in reliable communication tools is key.

Essential Communication Tools

Ham Radios (e.g., Baofeng UV-5R)

A crucial communication tool that allows you to send and receive messages across long distances without relying on the internet or cell networks.

Emergency Crank Radios

Solar and hand-cranked radios like the Eton American Red Cross FRX5 or Midland WR400 can receive weather alerts and emergency broadcasts.

Two-Way Radios

Compact and effective for short-distance communication, with models available from brands like Motorola or Cobra.

Satellite Phones (e.g., Iridium 9575, Globalstar)

A reliable backup communication tool for when cell networks go down.

6. Local and Global Alerts and Warning Systems

Stay informed during emergencies with various alert systems, apps, and services that provide crucial information about crises in your area.

Essential Alert Systems and Services

Wireless Emergency Alerts (WEA)

A free service that sends out emergency notifications via text message to your cell phone.

FEMA App

The official app for emergency alerts, weather updates, and disaster preparedness advice from the Federal Emergency Management Agency.

Alert Ready

A national public alert system for broadcasting urgent warnings, weather alerts, and emergency information in Canada.

NIXLE

Provides local government alerts and emergency notifications across a wide range of topics.

NOAA Weather Radio

A reliable source for receiving weather alerts, warnings, and real-time emergency information across the U.S.

7. Off-Grid Energy Providers and Resources

Securing an off-grid energy source is vital to maintaining your self-sufficiency and preparedness plan when the grid goes down. These resources can provide backup power and sustainable energy solutions.

Off-Grid Energy Providers

Goal Zero

Specializes in solar-powered generators, solar panels, and battery systems for off-grid energy.

Renogy

Offers solar panels, solar kits, and other off-grid energy products.

Windy Nation

Known for providing wind and solar power kits, charge controllers, and backup battery systems.

Blue Sky Energy

Offers solar controllers and off-grid power products ideal for prepping setups.

8. Sustainable Living and Homesteading Resources

Sustaining yourself and your family during an emergency involves more than just stockpiling supplies; it's about creating a sustainable lifestyle that provides for your needs. These resources can guide you in learning how to homestead, grow food, and be self-sufficient.

Sustainable Living Resources

Mother Earth News

A magazine and website covering everything from organic farming and gardening to renewable energy and home skills.

Permaculture Institute

Provides resources and courses on creating self-sustaining ecosystems and sustainable food production systems.

The Self-Sufficient Life

A website focused on homesteading skills such as gardening, animal husbandry, food preservation, and off-grid living.

Homesteading Forum

A community-driven platform offering advice, resources, and knowledge sharing for homesteaders and preppers.

The Permaculture Research Institute

A hub for permaculture knowledge, providing courses and resources on sustainable living and regenerative practices.

This collection of resource lists is meant to equip preppers with the knowledge, tools, and supplies they need to build a successful and resilient preparedness plan. By utilizing these resources, you'll have access to expert advice, gear, and communities that can help you stay safe, self-sufficient, and prepared for any disaster or crisis scenario.

DIY Templates for Home Projects

Creating a secure, self-sufficient, and resilient home environment during emergencies often involves a variety of hands-on projects. The following DIY templates for home projects are designed to help preppers, homeowners, and survivalists build effective solutions for improving home security, energy efficiency, food storage, and overall preparedness. These simple and practical projects can be constructed using basic materials and tools, and most are designed to be scalable based on individual needs.

1. DIY Emergency Shelter (Temporary or Semi-Permanent)

An emergency shelter is vital when dealing with natural disasters or when you're forced to evacuate your home. A DIY emergency shelter can be created quickly using minimal resources.

Materials Needed:

- Tarps or emergency Mylar blankets
- Rope or paracord
- Tent poles, wooden dowels, or bamboo sticks
- Ground tarp (for moisture barrier)
- Duct tape or waterproof adhesive

Instructions:

Select the Location: Choose a safe, level area with minimal risk of flooding or falling debris.

Create a Frame: Using your tent poles, wooden dowels, or bamboo sticks, build a simple frame for your shelter. A-frame or lean-to shapes are ideal for temporary shelters.

Attach the Covering: Stretch a tarp or Mylar blanket over the frame, securing the corners with rope or duct tape. Ensure it is taut to reduce wind resistance.

Waterproof the Ground: Lay a ground tarp under your shelter to prevent moisture from soaking through. Consider layering extra tarps for added insulation if necessary.

2. DIY Rocket Stove for Cooking

A rocket stove is an efficient and portable way to cook without electricity or gas. It burns small twigs, branches, and other biomass materials, making it ideal for off-grid cooking during emergencies.

- Materials Needed:
- A metal can or steel pipe (for stove body)
- Small steel pipe (for chimney)
- Tin snips or hacksaw
- Drill with metal bit
- High-temperature stove cement (optional)
- Fireproof bricks or mortar (for base)

Instructions:

Prepare the Base: Set up a fireproof base using bricks or stone. This will help prevent fire hazards and keep the stove steady.

Create the Stove Body: Cut a large metal can or steel pipe to create the body of the rocket stove. Use tin snips or a hacksaw.

Add a Chimney: Drill a hole at the top for a small steel pipe to serve as the chimney. The chimney should angle slightly to allow smoke to escape.

Assemble the Stove: Attach the chimney pipe to the stove body. Use high-temperature stove cement to seal any gaps and ensure efficient heat retention.

Fuel and Burn: Place small twigs and branches in the combustion chamber. The stove should create a focused flame that heats the cooking surface.

3. DIY Water Filtration System

When clean water sources are unavailable, a DIY water filtration system can help ensure your drinking water is safe. This system can be built using common materials that purify and filter water.

Materials Needed:

- Large plastic bottle (or bucket)
- Gravel (small size)
- Sand (medium size)
- Activated charcoal
- Coffee filters or cheesecloth
- A knife or scissors

Instructions:

Cut the Bottle: Cut the bottom off the plastic bottle to create a filtration chamber.

Create the Filter Layers: Layer the filtration materials inside the bottle in this order:

At the bottom, place a layer of gravel for coarse filtration.

Above the gravel, add a layer of sand for finer filtration.

On top of the sand, place activated charcoal for absorbing chemicals and improving taste.

Prepare the Filter: Cover the top of the bottle with a coffee filter or cheesecloth to prevent particles from entering the water.

Pour Water Through the Filter: Pour dirty water into the top and let it pass through the layers. Collect the filtered water in a container beneath the bottle.

4. DIY Solar Oven for Cooking

A solar oven harnesses the power of the sun to cook food, making it a great off-grid solution during power outages or survival situations.

Materials Needed:

- Cardboard box (large enough for your pots)
- Aluminum foil
- Black paint
- Clear plastic or glass (for a lid)
- Insulation material (styrofoam, cotton, or newspaper)
- Scissors and tape

Instructions:

Prepare the Box: Take a large cardboard box and cut a flap on the top that can be angled towards the sun.

Add Reflective Material: Line the inside of the box with aluminum foil, ensuring that the shiny side is facing up. This will help reflect and concentrate the sunlight into the cooking chamber.

Paint the Bottom Black: Paint the bottom of the box with black paint to absorb heat more effectively.

Insulate the Box: Line the sides and bottom with insulation material (like styrofoam or layers of newspaper) to retain heat.

Add the Lid: Cover the top with clear plastic or glass to trap the heat inside, creating a greenhouse effect.

Cook Using the Sun: Place your food in pots or trays inside the solar oven, positioning the reflective flap toward the sun. Cooking times will vary depending on the sunlight and temperature.

5. DIY Emergency Firestarter Kit

A firestarter kit can help you start a fire quickly during an emergency, whether you're camping, in the wilderness, or at home without power.

Materials Needed:

- Cotton balls or dryer lint
- Vaseline or petroleum jelly
- Small tin can (e.g., Altoids tin)
- Wax (optional)
- Matches or a lighter

Instructions:

Prepare the Cotton Balls: Take cotton balls or dryer lint and coat them with a generous amount of petroleum jelly. This creates a firestarter that burns for a longer period.

Store in a Container: Place the prepared cotton balls or lint inside a small tin can or container. If you have beeswax or candle wax, you can melt it and pour it over the cotton balls for additional waterproofing.

Use the Firestarter: When you're ready to start a fire, remove one of the cotton balls from the container and strike a match. The cotton ball should catch fire easily and provide a long-lasting flame for your fire.

6. DIY Vegetable Garden Planter Box

For those looking to grow their own food in a small space, a vegetable garden planter box is an easy project that can be set up quickly in your backyard or on a balcony.

Materials Needed:

- Wood planks (e.g., cedar or pine)
- Nails or screws
- Hammer or drill
- Saw (if wood needs to be cut to size)
- Landscaping fabric (optional)

Instructions:

Cut the Wood: Use a saw to cut the wood planks to the desired size for the planter box. A typical size is 4 feet long, 2 feet wide, and 1 foot deep.

Assemble the Box: Attach the sides of the planter box using nails or screws. Ensure the corners are square and secure.

Add Drainage: Drill small holes in the bottom of the box for proper drainage.

Prepare the Soil: Line the bottom of the box with landscaping fabric to prevent soil from escaping through the drainage holes. Fill the box with soil and compost to create a nutrient-rich environment for your plants.

Plant Vegetables: Choose vegetables suited to your climate and planting zone. Plant your seeds or seedlings in the prepared box, following the recommended spacing guidelines.

7. DIY Solar-Powered Battery Charger

A solar-powered battery charger can be a lifesaver when you need to power essential devices during a grid-down scenario. This charger can be built using solar panels and a few simple components.

Materials Needed:

- Small solar panel (5W or higher)
- Charge controller
- 12V battery
- Diode (to prevent reverse current)
- Charging cable and connectors

Instructions:

Connect the Solar Panel: Attach the solar panel to the charge controller, making sure to connect the positive and negative terminals correctly.

Attach the Battery: Connect the charge controller to the 12V battery, ensuring proper polarity.

Add the Diode: Place the diode in line with the positive cable to prevent reverse current.

Test the Charger: Place the solar panel in a sunny location, and check the battery for charging status. Use the battery to power essential devices like radios or lights.

These DIY templates for home projects can provide you with practical solutions to improve your home's self-sufficiency, security, and preparedness. Each project is customizable based on your available resources, skills, and needs, and they will help ensure that your home is ready for any crisis scenario.

Emergency Plan Examples

Creating an effective emergency plan is crucial for ensuring safety, organization, and peace of mind during a disaster or crisis. Below are several examples of emergency plans designed for different situations, from natural disasters to civil unrest, providing a solid framework for preparedness.

1. Family Emergency Plan Example

A family emergency plan helps ensure that every family member knows what to do, where to go, and how to communicate in case of an emergency.

Components of the Plan:

- **Contact Information:** List of emergency contacts, including family members, neighbors, local authorities, and out-of-state contacts.
- **Evacuation Routes**: Map of your home and surrounding areas showing multiple routes to leave, including safe meeting points.
- **Emergency Kit:** A list of essential items to include in your emergency kit (water, food, first aid, flashlight, batteries, etc.).
- **Shelter:** Pre-designated emergency shelters or safe locations to go to (e.g., relative's house, community center, etc.).
- **Communication Plan:** Designate an out-of-state family member as the primary contact for checking in and providing updates. Ensure everyone has the phone numbers memorized or saved in their devices.
- **Special Needs:** Make provisions for family members with special needs (elderly, infants, pets, or anyone who requires medical attention).
- **Pets:** Have a plan for pets, including food, water, leashes, carriers, and any medication they may need.

Example:

In case of an earthquake, your family's plan might involve meeting at a local park if the house becomes unsafe. The plan includes checking in with your out-of-state contact, and having a family member responsible for taking care of the pets.

2. Home Emergency Evacuation Plan Example

An evacuation plan is essential when your home is in danger due to natural disasters, fire, or other hazards.

Components of the Plan:

- **Evacuation Routes:** Clearly marked exits for every room in the house. Identify at least two routes from each room to ensure safe egress in case of fire or blockage.
- **Safe Room:** A designated room with minimal windows, sturdy doors, and supplies in case evacuation is not possible.
- **Emergency Contacts:** List local emergency services and nearby relatives or friends who can provide shelter or help.
- **Assembly Point:** A specific location outside the home (e.g., a neighbor's house, a community center) where everyone meets after evacuation.
- **Supplies Checklist:** A list of items to take when evacuating, including documents, clothing, first aid kits, medications, and emergency food/water.

Example:

If there's a fire in the neighborhood, your evacuation plan might include gathering at a neighbor's house, which has a designated emergency kit. Your children know to go directly to the neighbor's house without stopping to collect personal items.

3. Disaster Response Plan Example (Natural Disaster)

A disaster response plan is designed to ensure preparedness for natural disasters like hurricanes, earthquakes, tornadoes, or floods.

Components of the Plan:

- **Pre-Disaster Preparation:** Ensure the home is fortified (e.g., storm shutters, roof inspections), and all family members have essential items ready for evacuation.
- **Alert System:** Sign up for emergency alerts via local agencies (e.g., NOAA, FEMA). Know how to receive and interpret emergency alerts and warnings.
- **Evacuation Instructions:** Detailed routes to evacuate and the nearest shelters, taking into account roadblocks and traffic conditions.
- **Communication Plan:** Establish check-in times for family members and assign a meeting location outside the affected area.
- **Community Resources:** Know local resources such as Red Cross shelters, food banks, and emergency clinics.

Example:

In preparation for a hurricane, your disaster plan includes boarding up windows, securing the family's emergency kit, and deciding to head to an inland shelter. Family members know to bring important documents and their emergency medicines.

4. Pandemic Preparedness Plan Example

A pandemic preparedness plan helps families manage the specific challenges that arise during a health crisis, such as widespread illness.

Components of the Plan:

- **Health Precautions:** Regularly wash hands, wear face masks when needed, and maintain social distancing when possible. Keep a thermometer and fever-reducing medicine available.
- **Medical Supplies:** A list of medications, sanitizers, and other medical supplies needed for each family member, including prescription refills.
- **Communication with Healthcare Providers:** A contact list for healthcare providers, pharmacies, and local health department resources.
- **Remote Work and School:** Identify resources for remote work and school (laptops, internet, etc.), as well as backup childcare if necessary.
- **Food and Water Supplies:** Ensure a sufficient stockpile of food, water, and personal hygiene products in case of quarantine or supply chain disruptions.

Example:

In the event of a pandemic, your plan includes staying indoors for 14 days, avoiding large gatherings, and utilizing telemedicine services for non-emergency health concerns. You will also rotate through your food storage to avoid running out of essentials.

5. Civil Unrest and Riots Emergency Plan Example

In case of civil unrest or riots, having a plan for staying safe and managing personal security is critical.

Components of the Plan:

- **Safe Locations:** Identify locations inside the home to take shelter, such as interior rooms away from windows, or a basement if necessary.
- **Evacuation Routes:** Plan alternate routes to avoid areas of conflict, including less-traveled back streets and trails.
- **Self-Defense Measures:** Secure tools for self-defense (e.g., pepper spray, tactical flashlight, or emergency whistle) and ensure all family members know how to use them.
- **Communication:** Designate a family member who is not in the immediate area to act as a point of contact and update the group on any developments.
- **Secure Your Property:** Fortify doors, windows, and other entry points. Reinforce the home with barricades if possible.

Example:

During an outbreak of civil unrest, your family plan may involve staying inside with all doors locked and barricaded. Your plan includes staying off social media to avoid disinformation and relying on news sources for real-time updates.

6. Financial Emergency Plan Example

A financial emergency plan ensures that your finances are prepared for economic hardship, such as job loss, economic collapse, or a sudden large-scale crisis.

Components of the Plan:

- **Emergency Fund:** Have at least 3-6 months' worth of living expenses saved in an accessible account.
- **Diversified Assets:** Invest in diverse assets (e.g., cash, gold, cryptocurrency, and real estate) to safeguard against currency devaluation or banking disruptions.
- **Debt Management:** Create a plan to reduce high-interest debt and avoid accumulating additional liabilities during an economic crisis.
- **Income Stream Backup:** Identify potential side jobs, passive income, or work-from-home opportunities to generate income if your primary income source is disrupted.
- **Financial Assistance Resources:** Keep track of government aid programs, nonprofit organizations, and local community groups that may provide financial assistance during a crisis.

Example:

If there is a sudden economic downturn, your financial plan includes using emergency savings to cover necessary expenses, while exploring additional income sources like freelance work or selling unused items to generate cash.

7. Community Network Plan Example

A community network plan ensures that you can rely on neighbors, friends, and other local contacts during a crisis for mutual aid and shared resources.

Components of the Plan:

Neighbor Coordination: Organize with nearby neighbors for shared resources (e.g., food, water, security) and mutual assistance.

Group Communication: Establish a method of communicating with your community network during power outages (e.g., walkie-talkies, ham radio, or pre-established signal systems).

Shared Resources: Pool resources for bulk buying (food, water, medical supplies) or communal gardening.

Community Tasks: Assign roles and responsibilities to neighbors to ensure coverage in key areas like security, child care, or elderly assistance.

Example:

In an extended power outage, your neighborhood plan involves bartering with neighbors to share cooking or water filtration resources. You and your neighbors are also aware of emergency routes and rallying points in case evacuation is necessary.

Emergency plans must be tailored to your specific needs, environment, and the risks you face. Whether it's a family evacuation plan, a financial emergency plan, or a disaster-specific response plan, having a clear, organized strategy for different scenarios helps reduce panic and increases the likelihood of a successful outcome in times of crisis. Be sure to review and practice these plans regularly to keep everyone prepared and informed.

Made in the USA
Monee, IL
13 January 2025

76753808R00109